THRIVING MINDS

in the Age of AI

THRIVING MINDS

in the Age of AI

The Brain Science of Thinking Clearly and Living Well

Professor Selena Bartlett

First published 2026
Phineline Publishing
Brisbane, Queensland 4010 Australia

All enquiries should be directed to the author.
ISBN 978-0-9990997-7-3 (paperback)
ISBN 978-0-9990997-8-0 (ePub)

When machines increase our power,
We must increase our collective wisdom.

Contents

Introduction

The Question That Changed Everything

It was 1989. I was twenty-four years old, standing in a psychiatric ward in Brisbane, and nothing in my pharmacy training had prepared me for what I was about to see.

My sister Francesca sat rigidly in a dim room, her body taut, her eyes far away, my hands clenched by my side, watching. This was my sister, the charismatic, funny, and open person I had grown up with. And now she was here, in a place that appeared more traumatic than the treatment.

One question formed in my mind with a strength that never fully left me.

Why? How did we get here? And could it have been different?

The doctors had given Francesca's condition a name that revealed just how little they understood it, a schizophreniform-like illness, a label of last resort, arrived at by eliminating everything else. The treatment which followed was, in many ways, more damaging than the diagnosis. Heavy sedation. Isolation. An endless cycling of medications, hospitalisation, where each took something from her: a friendship, a dream, a part of herself that never returned.

By 2005, she was on clozapine, the most potent antipsychotic available, and suffering its consequences: seizures, metabolic disease, and diabetes. She was no longer quite Francesca; she had become, as happens to so many people in that system, her illness.

Standing in that ward in 1989, I did not know that this moment might be the next three decades of my career. I could not save Francesca; however, the question she planted in me never stopped driving me forward.

The Science I Went Looking For

I had a clear path mapped out before that day, a career as a community pharmacist, just like my parents, who had run their pharmacy in Nanango for forty years. My mother always said it was a good job for a girl: steady, respectable, something you could step back from when children arrived. But I had become a pharmacist who understood the mechanisms of drugs without understanding the minds they were meant to help. I knew pharmacokinetics, pharmacology, and medicines. I did not know why some people developed mental illness, and others did not, why one treatment worked brilliantly for one patient and catastrophically for another, or what had happened to my sister.

At twenty-six, I did something that looked reckless to everyone who knew me. I left pharmacy and enrolled in a PhD in neuropharmacology at The University of Queensland, followed by a postdoctoral fellowship in developmental neurobiology at the Australian National University. I was married and walking into a male-dominated world I did not yet fully understand, with no map and no guarantee of arrival anywhere.

What I found was extraordinary: new brain imaging technology was lighting up brain pathways that had never been visible before. I studied addiction and the brain's reward system, examining how substances hijack the same circuits that evolved to guide us toward food, water, and connection. I developed medications targeting these systems, and from the outside, it looked like exactly the right path.

The more I came face to face with what we still did not understand. I could see the extraordinary elegance of neuroscience, the circuits, the molecules, the precision, and yet I felt the painful distance between that knowledge and the blunt reality of the sister's suffering. In 2001, I moved to San Francisco for what became an extraordinary period of my career, working alongside some of the top scientists in the field. Immersed in world-class research, my thinking became firmly anchored in the medical model, mapping brain circuits, refining targets, and developing better medicines. Then, in 2009, something changed: a seed was planted that would always change the direction of my thinking and my life.

The Discovery That Changed the Question

The moment that changed everything did not come from my own research. It came from a colleague's lab, which was researching neuroplasticity, and one afternoon she showed me something I was not prepared for: an adult brain, visibly and measurably transforming in real time. I had encountered the concept of adult neuroplasticity for the first time. The adult brain was not fixed; given the right conditions, it could change. This also means the patterns formed by experience, stress, and trauma were not permanent!

This changed everything. I had spent years studying the brain, yet I had never truly turned that lens on myself. I was running on adrenaline, driven by achievement, work, and the demands of raising a family. My lab was funded, my career was advancing, and from our home in the Berkeley Hills, the view was so beautiful our friends called it Camelot.

But beneath the surface, I was under constant, unrecognised stress and starting to unravel. Francesca had already passed in 2006, before I understood neuroplasticity, and that realisation carries a weight I still feel. What I learned came too late to help her.

Yet discovering neuroplasticity reshaped everything that followed. It gave me a new way to understand what she had been going through and what I had missed. It also gave me a way to change how I respond to stress, how I show up for others, and how I make sense of loss.

What I Finally Understood About My Sister

As I involved myself in the science of early life experiences, adverse childhood experiences, trauma and neuroplasticity research, I began to see Francesca's story differently. Not as a mystery of only faulty genetics and neurotransmitters, but as the outcome of neurobiology meeting circumstance at the most vulnerable moment of brain and child development.

I was born three years before her, when our family lived with my grandmother in Brisbane. My grandmother held me, rocked me, and called me "Seen", a name she used only for me. I arrived in a caring hug and was given attention, love and presence, exactly what developing brains require.

Francesca was born into different circumstances. My parents had relocated to Nanango to build their pharmacy from nothing. They were exhausted, overstretched, and far from family support. Within a few months of her birth, my mother was pregnant again. Francesca was desperately loved. However, the quality of early emotional attention and the consistent serve-and-return interactions that wire healthy brain architecture during the first three years of life were harder to sustain in those conditions.

Her brain adapted to what was available; it was neither her fault nor my parents'. Francesca died before I could tell her what I had come to know. But her story became the foundation of everything I now teach, research, and believe about the brain's ability for both damage and repair.

Neuroplasticity is a key idea in this book. Simply put, it means the brain can change and modify based on what we do and experience.

This ability is not just for children; our brains can adapt and rebuild at any age.

The brain is not something that stays the same and just gets older. Instead, it is always changing and is formed by how we live. Every habit we build, every way of thinking we repeat, and every place we spend time in changes our brains. The more we repeat something, the stronger those brain pathways get, and if we stop using them, they fade.

This goes both ways; brains that deal with constant stress, scattered attention, or a constant state of being on edge become more reactive and less stable. Brains that get regular movement, good sleep, focus, and real, safe and secure interpersonal connections become stronger, clearer, and better able to handle tough times.

Neuroplasticity is not simply a scientific idea; it's how we change. The habits and environments we create for ourselves and others are not just small details. They shape the kind of brain we end up with. I tried a simple daily habit from a book I had previously ignored. I asked myself tough questions each morning, changed my usual ways of thinking, and became more aware of what I was putting into my mind. The changes happened slowly at first, then became clear. What I was really doing, even before I knew the science, was changing the way my brain worked. The brain I have now is not the same as the one I had then. The pace of change required in the age of AI is unlike anything I have experienced before. Through my work as a neuroscientist studying AI and mental health, I could see how quickly these tools were beginning to shape attention, behaviour, and decision-making. In our universities and schools we watch students integrate generative AI into how they think, learn, and solve problems. At home, I saw it through my son working inside an AI start-up, and in my children as they tried to understand what their future might look like in a world where the rules of work were shifting.

Why This Matters Right Now

Over the past two years, something has shifted in the invitations I receive from law firms, Universities, hospitals, government departments, and corporate boards, which are asking questions that are all variations on the same theme: What does artificial intelligence mean for us? How do we adapt without losing ourselves in the process?

AI tools, unlike others in the past, have direct consequences for our social identity, education, and the meaning of expertise have landed all at once. In universities, we are asking how to teach critical thinking when machines can generate essays. In law firms, partners are asking what happens to junior lawyers whose learning has traditionally come from the slow, effortful work of research and drafting. In hospitals, clinicians are asking what it means to practise medicine when diagnostic AI begins to outperform human pattern recognition. These are not abstract questions; they are becoming daily realities.

Beneath the excitement of what AI can do, there is a quieter and more important question: what does this pace of change do to the human nervous system? What happens to identity, to purpose, to psychological stability when the cognitive skills that define a person can be replicated by a machine?

The answer, from neuroscience, is that uncertainty at this scale activates the same neural circuitry as physical threat. The brain registers an unknown future as danger. Stress rises, our attention narrows, reactivity increases, and we find it harder to collaborate. And when that sustained pressure lands on a nervous system already strained by years of digital overstimulation, the effect compounds.

This is the moment I have spent my career preparing to address because the underlying question, how human brains adapt to accelerating pressure without failing, is the same question Francesca's story first asked me in 1989.

The answer, then as now, begins with understanding what the brain actually needs to change, and then providing it.

What This Book Will Give You

Thriving Minds in the Age of AI is not a book about technology. It is a book about the human body, brain, and mind navigating an AI world, and what it takes to do so well. If you read this book, you will walk away with a practical roadmap for protecting and strengthening your mind in the face of exponential change, so you can adapt, perform, and thrive no matter how quickly AI transforms your field.

This is not a rigid prescription, but one example of how to weave movement, attention management, and recovery into an ordinary workday. Even one or two changes can make a measurable difference, and the routines that last are the ones that fit your life. If your days do not follow a typical office schedule, because you work in shifts, travel frequently, or have caregiving responsibilities, adapt these principles to suit your reality. For shift workers, movement and recovery might occur before or after unconventional work hours; for caregivers, micro-breaks or shared moments of movement might be more practical than long blocks of time; for frequent travellers, maintaining focus on the go or taking short walks in airports can keep routines alive. The crucial point is flexibility: these foundations are meant to support you, whatever shape your days take.

If you want to begin right now, try a single five-minute walk before you start your workday or put your phone aside for the first ten minutes after waking. These quick-start actions require no preparation, yet they immediately begin building new pathways for attention and resilience. Start with just one today and notice what changes.

Each chapter addresses one dimension of the research on building a resilient and flexible mind. We examine what genuine resilience looks like at a neural level, not grit or denial, but the specific conditions under which brains recover and grow. We look at the fundamentals: sleep, movement, food, and the quality of human interpersonal connections, often dismissed as wellness clichés, but in fact, the biological infrastructure without which no amount of strategy or skill development will hold. We move through the mechanics of attention and how to reclaim it from systems designed to capture it.

We honestly examine ourselves, the biases and blind spots that shape professional decision-making, and what it means to build daily habits that gradually and measurably rewire the brain toward thriving rather than merely surviving. The final chapters turn toward the future. Not with fear, but with the orientation that three decades of neuroscience research has given me: the brain that built civilisation, that formed the relationships and exercised the ethical judgement and held the grief and made the breakthroughs, that brain is not obsolete. It is, in the most important respects, irreplaceable.

What AI cannot do is be human. It cannot co-regulate another person's nervous system with its presence. It cannot hold someone's joy and grief or exercise moral and ethical courage. Those capacities are still human, biological, relational, and, with the right conditions, endlessly renewable.

That is what this book is about. Not surviving the AI moment but thriving in it by understanding, at last, what the human brain is genuinely for.

Professor Selena Bartlett

Brisbane, 2026

CHAPTER ONE

The World Changed

Why AI is rewriting the rules faster than we realise

Why AI is rewriting the rules faster than we realise

It is three in the morning. I am lying in the dark, wide awake, and my mind will not stop.

Earlier that evening, I had been listening to Mo Gawdat, former chief business officer of Google X, talk about artificial general intelligence. He was calm, measured, almost mathematically precise, and something transformed in me as I thought about what the careers of my children, students, and colleagues will look like in a decade. What happens to professional identities when the analytical work that defines a career, such as the synthesis, the drafting, the pattern recognition, can be done in seconds by a machine?

I did not feel afraid, exactly, more an unease, as if I was standing by the edge, seeing something huge below. That feeling has a name in neuroscience: it is the stress response firing in the presence of uncertainty. The brain is a prediction machine, and when the future becomes uncertain, it registers the unknown as a threat. What I experienced that night was not irrational; it was understandable and neurobiological. And if you work in law, medicine, education, or any profession where thinking is the currency of your career, you have likely felt some version of it too.

This book is not about whether AI will transform professional life. It will, and it already is. It is about understanding the processes happening in the brains and minds of the people navigating that transformation, and why the human capacities we have long undervalued are about to become the most important ones we have.

THE SPEED OF CHANGE

Most of us understand exponential growth as a mathematical concept. Fewer of us have experienced it firsthand. We are feeling it now. Consider a single example. For decades, determining the three-dimensional structure of a protein, the precise shape that governs how it interacts with other molecules, how drugs bind to it, and how diseases exploit it has required years of painstaking laboratory work. X-ray crystallography and cryo-electron microscopy, and a PhD student or students might spend the better part of their doctorate solving the structure of a single protein. The global database of known structures, built over half a century of scientific effort, held around 170,000 entries. DeepMind released AlphaFold2, which predicted the structures of virtually every known human protein. With AI, DeepMind predicted over 200 million protein structures in just weeks. In simple terms, what once took decades or couldn't be done at all can now be done almost instantly, dramatically accelerating how we understand disease and develop new treatments.

What had taken the combined effort of the global scientific community 50 years to achieve, AI replicated in years, and now in months. This breakthrough was recognised with the Nobel Prize in Chemistry, awarded to Demis Hassabis and John Jumper for their work on AlphaFold.

For drug discovery, this is transformative because researchers can now explore how mutations alter protein shape, identify how diseases exploit structural vulnerabilities, and design molecules to intervene all before setting foot in a laboratory. The early stages of drug development, once measured in years, can now be compressed in weeks. As a lawyer, you might be thinking that it is science, not law. But the shift is identical; legal research that once took hours now takes seconds. Contract analysis that required a senior associate now runs in the background. The due diligence that consumed teams for weeks is being compressed. This is what exponential actually feels like from the inside: not a dramatic rupture, but a quiet, relentless acceleration, until one day you look up and the landscape has changed entirely. Before we dive into the details of how AI is transforming our professional lives. I invite you to pause and ask yourself:

Where have you noticed unexpected acceleration in your own world?

Is there an area of your work or daily life where the pace seems to outstrip your ability to adapt? Hold that example as you read, because understanding how exponential change feels, not just intellectually but personally, is where the science becomes most useful.

To help you translate these ideas into your own context, consider the following prompts:

Where is uncertainty highest in your week?

Are there aspects of your role or life where the rate of change leaves you feeling stretched?

Which routines or assumptions have you had to question, and how did you respond?

What the Digital Environment Is Doing to Our Minds

AI has arrived at a time when our brains are already under pressure from a decade of digital acceleration, and that matters enormously for how we respond.

The human brain runs on attention; whatever is repeatedly focused on is wired into the neural architecture. This is neuroplasticity at its most fundamental level: the brain changes in response to what it repeatedly encounters. Social media platforms understood this before neuroscientists had fully articulated it. Notifications, infinite scroll, variable reward schedules, and algorithmic curation of feeds are precisely designed to activate dopamine and neurochemical pathways that evolved to help us find food and avoid predators. Now those same pathways fire in response to a like, a comment, a negative news headline.

The result, for many professionals, is a nervous system that rarely fully settles. Not dramatic stress, as most people would not describe themselves as acutely anxious, but persistent low-grade stress. Patience is thinner than it used to be, and sleep is slightly worse than it should be.

Take Sarah, a mid-career lawyer who wakes up already thinking through her day's first meeting. She checks her phone before getting out of bed, quickly scrolling past twenty emails and half a dozen notifications from colleagues and clients who expect a reply before she arrives at the office. Throughout the day, her attention is constantly pulled between drafting briefs, responding to messages, and processing a stream of updates from multiple platforms. By late afternoon, she realises she has skipped her lunch break and cannot recall the last time her mind felt fully quiet. Even when she returns home, her mind is buzzing with half-finished thoughts, and her sleep, when it arrives, is shallow and unrestful. Sarah is not gripped by panic, but her nervous system is never truly at rest. This is the modern professional baseline: not crisis, but an ongoing undercurrent of tension that makes patience thin and real restoration elusive.

Add to this the stress physiology of constant social evaluation. The brain does not cleanly distinguish between physical and social threats. Exclusion, humiliation, and professional comparison activate neural circuitry that substantially overlaps with that activated by physical danger. For professionals whose status and identity are closely tied to their expertise, the prospect of its automation is not simply a concern. It registers, at some level, as a threat to survival.

The stress many professionals feel about AI is not a weakness. It is the nervous system doing exactly what it was designed to do, scanning for danger in conditions of rapid change. There is something else being eroded that receives less attention. One of the more striking findings in modern neuroscience is that human brains regulate each other. When two people are genuinely present with one another, emotionally connected and focused, their neural activity begins to synchronise, a phenomenon known as inter-brain synchrony. It is a measurable physiological process that underpins learning, trust, and emotional strength and can connect us across distances. When a client is anxious in your consulting room, your calm nervous system can literally help regulate theirs. That process is biological and cannot be fully replicated by a video call, let alone by a chatbot. It is one of the most powerful things a skilled professional offers, and it is almost never mentioned in capability frameworks.

Learned Helplessness in a Digital World

Around the time I began thinking seriously about these questions, I wrote a newsletter article on learned helplessness, a concept in behavioural neuroscience first described by Martin Seligman and Steven Maier in the 1960s. When people are repeatedly exposed to stressors they cannot control, they eventually stop trying to change their situation. The motivation system dampens, their agency erodes, and the person begins to conserve energy rather than seeking solutions. For example, in workplaces where effort feels disconnected from outcome, in communities where system-wide barriers are immovable, and, increasingly, in the digital environment.

Experienced practitioners who have watched automation reshape industries have spent years mastering it. Citizens are overwhelmed by global crises delivered in real time, every day, to a screen in their pocket. When change appears both threatening and uncontrollable, passivity can quietly take root, not as a choice but as a neurological adaptation, and slide into learned helplessness.

This is the version of the AI moment that worries me most. Not the striking fiction scenario of machines replacing humans, but the quieter, more insidious risk that people begin to feel that nothing they do matters and that the wave is too large, the change too fast, the future too opaque, and they stop actively, intentionally shaping their response to it. The cycle of learned helplessness can only be broken when we remember that even the smallest action, piloting a tiny experiment, adjusting one habit, testing one new approach, puts agency back in our hands. You do not need a master plan to regain direction; you only need to start with one step you can control today.

But here is the critical insight that steadied me at three in the morning: learned helplessness is learned. And what is learned can be unlearned.

The brain that adapts to fragmentation could also adapt to focus. The professional who has outsourced certain cognitive tasks to AI can redirect that freed capacity toward precisely the human faculties that cannot be reproduced: ethical judgement, wisdom, and the kind of contextual insight that only comes from lived experience.

The Human Capacities That Are About to Become More Valuable

There is a version of the AI conversation that frames this moment as a zero-sum contest between human and artificial intelligence. That framing misunderstands what is happening, and it obscures the real risks of treating AI as a neutral partner in decision-making. The urgency of human judgement becomes clear when we consider how AI systems can amplify inequities. For example, facial recognition technologies have repeatedly been shown to misidentify people of colour at much higher rates than Caucasian individuals, leading to wrongful arrests and personal consequences. In healthcare, algorithmic tools used to prioritise patients have sometimes systematically underestimated the needs of Black patients, perpetuating inequities in access to care. These cases illustrate that behind the claim of efficiency lies a critical question: who is seen clearly, and who remains at risk of error. The stakes are not abstract: when professional expertise is replaced or sidelined without careful oversight, especially in law, medicine, or education, current biases in data can become encoded and scaled, with real and permanent impacts on people's lives.

AI increases cognitive throughput, but it does not yet increase wisdom. It can accelerate analysis, but cannot exercise the ethical judgement that determines what should be done with that analysis. It can synthesise information at scale but cannot read the situation, feel and sense a shift in a client's affect, sense when a negotiation is moving toward an irretrievable breakdown, or hold the space a grieving family needs before the paperwork begins.

These are not soft skills but our human capacities, built over years of experience and practice at home, in school, in our community, in the workplace, and in Society. And as AI increases cognitive speed, these capacities become scarcer and more valuable, not less.

Consider what this means concretely for professionals:

For lawyers: the capacity to read a witness, to sense what a client is not saying, to exercise the moral courage to tell someone what they do not want to hear, and these are not things a model can do. They are things a regulated, experienced human being does.

For educators: the ability to notice that a student is struggling before they say so, to adjust a lesson in real time in response to the energy in the room, to be the adult presence that a young person can anchor to during difficulty, and this is irreplaceable.

For health workers: the therapeutic relationship itself is part of the treatment. Presence, engagement, and the felt sense of being genuinely seen are not peripheral to clinical outcomes. In many contexts, they are central.

None of this means ignoring AI or resisting its adoption. The professionals who will thrive are those who can use AI to amplify their capacity while simultaneously deepening the distinctly human skills AI cannot touch.

Shaping the Brain, We Are Building

Neuroplasticity works in both directions. The brain adapts to fragmentation if that is what the environment repeatedly delivers. It also adapts to focus, depth, and intentional practice, if that is what we ask of it.

This is the argument that runs through this book. The digital environment is already reshaping the brains of humans, professionals, developing minds, and people trying to do good work under difficult conditions. AI is accelerating that reshaping, and the question is not whether our brains will change; it is whether we shape that change consciously or let it happen to us by default.

Thriving in an AI world requires something that no algorithm is able to do: a regulated nervous system. A nervous system that can tolerate uncertainty without defaulting to threat narratives. That can focus deliberately on an environment designed to fragment attention. That can remain present with another person in a moment, rather than retreating to the managed safety of a screen.

It requires real relationships, with the full bandwidth of your presence, because inter-brain synchrony is not a luxury and is a biological necessity for resilience and clear thinking. It requires the particular kind of professional courage that comes from knowing who you are and what you are here for, an identity robust enough to accommodate change without being destabilised by it.

Lying awake that night, I eventually realised the unending anxiety was a signal, and the brain's threat scanning in conditions of rapid change is doing exactly what it should. The question is what we do with that signal. We can let it tip into helplessness. Or we can let it catalyse the design of our habits, environments, professional practices, and sense of purpose, as the world has changed. The question before every professional reading these pages is the same one I faced in the dark: not whether the wave is real, but whether we learn to navigate it. The chapters that follow are about how.

CHAPTER TWO

Resilient Minds

We Become What We Repeatedly Do

Our mother, June, got up at 3:30 every morning.

Not occasionally and not when she felt motivated, not after a good night's sleep or a light day ahead. Every single morning of her adult life, she tied her walking shoes in the dark and walked for two hours through streets that were silent and still. Then she came home, ran the household, opened the pharmacy, and worked beside my father. Then she did it again the next day. As a neuroscientist, I now understand what my mother was doing each morning before dawn. With every run, she was raising her heart rate and releasing brain-derived neurotrophic factor (BDNF), the brain's key growth protein, along with exerkines, powerful signalling molecules from working muscles that travel to the brain to support repair, growth, and stress regulation.

Without knowing the science, she was training her nervous system to handle life's challenges. One dark morning at a time, she was building resilience. That is neuroplasticity in action.

What Resilience Actually Is

The word resilience has been so thoroughly absorbed into professional development culture that it has nearly lost its meaning. We speak of resilient leaders, resilient teams, resilient organisations, as though resilience were a trait some people possess, and others simply lack. Something you either have or you do not.

This is not what the science shows. Resilience is not a fixed characteristic, rather a neurobiological state, one that can be built, maintained, depleted, and rebuilt. It is the product of a nervous system that has been trained, gradually and repeatedly, to tolerate stress without tipping into dysregulation. It is the capacity to move through difficulty and return.

It is also important to recognise that the environment we live in matters. Many of the pressures professionals face are not simply the result of personal habits or choices. Workload expectations, economic insecurity, housing instability, and institutional inequalities all contribute to a baseline of stress that no individual can simply shake. When the wider environment is demanding or uncertain, such as when working hours stretch without compensation, when care responsibilities cut into rest, or when organisational cultures reward overwork and treat exhaustion as proof of commitment. These organisational factors produce a level of background stress that even the best personal strategies may struggle to fully offset. Not naming these realities risks leaving readers feeling solely responsible for what are, at least in part, collective challenges. If you see yourself in this struggle, know that resilience is never built in isolation from the systems around you.

There are concrete steps you can take to advocate for healthier organisational environments. Start conversations: Raise issues of workload, boundaries, or recovery practices in team meetings or with your manager. Participate in staff wellbeing, occupational health, or mental health committees within your company or institution to help shape supportive policies and practices. Advocate for changes supported by research, such as robust recovery periods, limits on after-hours email, or flexible scheduling. Use examples from respected guidelines or peer organisations to strengthen your case. Collaborate with colleagues to gather anonymous feedback or conduct pulse surveys on stress and workload, then present results to leadership to make the case for intervention. Engage with professional associations or unions to elevate systemic issues beyond your immediate workplace, thereby influencing sector-wide culture and standards. Lead by example by setting clear boundaries, taking breaks without apology, and encouraging others to do the same. Sometimes individual modelling opens the door to broader change.

Even a single conversation or policy suggestion can have ripple effects. By moving from private concern to shared initiative, professionals not only ease their own burden but also help to contribute to cultures that support everyone's resilience. By encouraging leaders and organisations to move beyond rhetoric and take concrete steps, we can create professional ecosystems where resilience is not an individual burden, but a shared foundation for everyone.

The Neuroscience of Doing Hard Things

Exercise is the most well-researched resilience intervention in neuroscience, and it works through pathways far beyond cardiovascular fitness.

When you move and genuinely, effortfully move, muscles release a cascade of signalling molecules collectively called exerkines. I spoke with Associate Professor Tara Walker about this on the Thriving Minds podcast, and she described what happens in the brain during and after exercise. Exerkines cross the blood-brain barrier and act directly on brain tissue. Exercise also stimulates the production of brain-derived neurotrophic factor, which supports the survival and growth of existing neurons and promotes the formation of new ones, particularly in the hippocampus, the structure most associated with memory, learning, and stress regulation.

The hippocampus is also one of the brain structures most vulnerable to chronic stress. Sustained stress exposure physically reduces its volume. Exercise, in a measurable and reproducible way, reverses this. But here is what I find most relevant to the professional context: exercise is controlled stress. You choose to raise your heart rate, push past comfort and then recover. And through that repeated cycle of activation and recovery, you are teaching your nervous system something that transfers far beyond the gym or the footpath: that difficulty is survivable, that your body can handle more than you thought, that you will come back from the edge.

Every hard run, every early morning walk, every set completed when you wanted to stop, these are not just physical achievements. They are lessons in stress tolerance that the brain applies everywhere.

This is why my mother's two-hour walks mattered; they were good for her heart, and she was training her nervous system, every single morning, in the foundational lesson of resilience: you can do hard things and survive them.

The same principle applies to cognitive and emotional challenge. Sitting with a difficult problem instead of immediately outsourcing it. Having a hard conversation instead of avoiding it. Tolerating the discomfort of not knowing, rather than reaching for the quickest available answer. These acts of deliberate difficulty build the same neural architecture as physical endurance. They demonstrate to the brain that strain is not a catastrophe.

What the Longest-Lived People Know

When I spoke with Dr Mario Martinez on the Thriving Minds podcast about his research with centenarians, that is, people who live past one hundred in good health, one finding stood out above all others.

Genetics account for approximately 20% of longevity. The remaining eighty per cent is shaped by our environment and beliefs.

The centenarians Dr Martinez has studied across multiple cultures share a remarkably consistent psychological profile. They believe they still have time and that ageing can be healthy. They see purpose ahead of them, not behind. They do not see themselves as finished. And that belief, that orientation toward future possibility, keeps them engaged in life in ways that are directly and measurably protective for the brain.

Engagement requires movement, cognitive effort, social connection, and meaning. The belief that there is more to come generates the behaviour that makes it possible.

My mother had this belief, expressed in her own direct way. She texted us two quotes so often that they became part of the family vocabulary. The first: "If you're going through hell, keep going", from Winston Churchill. The second: "work hard and everything follows". She did not mean these as inspiration. She meant them as instruction and a philosophy built from watching her own mother, from the decades in the pharmacy, from building security where none had been handed to her.

When I was ten years old, she told me, "My job is to make you financially independent." At the time, I heard a practical statement about money. What I understand now is that she was talking about something deeper than money. She was talking about capacity and about not being trapped by circumstance, about having the internal and external resources to make choices freely. She was describing what we would now call agency. And agency, as the research on learned helplessness makes clear, is one of the most powerful protective factors for brain health.

The belief that your efforts matter and that what you do changes what happens is not simply motivating. It is neurologically protective, and our brain and mind are engaged in living. It is the antidote to the quiet, corrosive passivity that chronic helplessness produces.

Resilience in the Age of AI

The professional world that readers of this book inhabit is undergoing a stress test unlike anything in recent memory. Not the acute stress of a single crisis, a market collapse, a pandemic, or a restructure, but the chronic, ambient stress of sustained uncertainty. The ground keeps shifting, and the skills that defined a career last year may be less central this year. The tools change faster than the training. The future remains, frustratingly, opaque.

This environment draws attention to the critical distinction between endurance and rest and stresses the importance of resilience. Endurance says: keep going regardless, push through and don't show the cost. This is the model my mother's generation inherited, and they built remarkable things with it. But the silent costs, which went unnoticed because they were never spoken about, were real and accumulated.

Resilience says something different. It says: build the capacity to sustain effort over time, which requires recovery. Develop the nervous system's ability to return to baseline after activation. Create the biological conditions under which clear thinking, creativity, and connection remain available, not just on good days, but under pressure.

In an AI-augmented workplace, the purely cognitive tasks, research, synthesis, first-draft generation, and pattern recognition at scale are increasingly being handled by machines. What remains distinctively human, and what AI cannot reproduce, is the capacity to function well under social and emotional complexity. To make judgements in conditions of uncertainty. To sustain creative thinking across time and to remain present with another person when the situation is difficult. These capacities are not soft and are the product of a deliberately built, repeatedly reinforced nervous system that has developed over time.

In a world where AI handles more of the cognitive effort, professional advantage shifts decisively toward the person with the most resilient nervous system, not the person with the most information.

This reframes the entire conversation about professional development. It is not primarily a question of what tools you can use. It is a question of what kind of brain you are building. What are you doing, daily and consistently, to train your stress response, to restore your nervous system's capacity to regulate, to maintain the biological conditions under which your distinctly human capabilities remain available?

These are not wellness questions but performance questions. They are the questions that, in the long run, determine who thrives, succeeds, and emerges, regardless of how sophisticated their AI tools are.

Resilience is not a trait you either have or don't. It is a biological capacity built through movement, recovery, belief, and the deliberate cultivation of the conditions in which a nervous system can thrive.

Building Resilience: What the Research Actually Recommends

We know, with considerable confidence, what builds a resilient nervous system. The challenge is not knowledge, it is implementation, particularly in professional environments that often reward the behaviours most corrosive to long-term resilience.

What follows is not a wellness list. It is a summary of the biological mechanisms that determine whether your brain can sustain high performance over time, grounded in the research and in the patterns I have watched shape generations of my own family.

Move Your Body, Every Day

The evidence for exercise as a direct brain intervention is now overwhelming. Two hours of walking every morning, as my mother did, is exceptional. But the research suggests that even 30 minutes of moderate aerobic activity, five days a week, produces measurable increases and sustained, long-lasting improvements in stress regulation, as well as significant reductions in stress reactivity.

The mechanism matters: you are not simply burning calories or improving cardiovascular fitness. You are releasing exerkines that support growth and repair. You are training the stress-response system through the controlled challenge of physical effort. You are repeatedly demonstrating to your nervous system that activation has an end.

Try tomorrow: Before breakfast, step outside, if you can, and walk for just five minutes. Set a timer if you need to. Notice how your mind and body feel during and after and treat it as a small experiment rather than a commitment. If five minutes feels too much, try two. The goal is not to achieve perfection, but to create a moment of movement you can repeat. Early wins build momentum; what matters is starting.

In practice: Find the form of movement you will actually sustain. Consistency matters far more than intensity. The brain does not care about the type of exercise, it cares about the regularity.

Distinguish Useful Stress from Chronic Noise

Not all stress is equal. Acute, bounded stress, such as a difficult deadline, a challenging presentation, or a demanding negotiation, activates the stress response in ways that, when followed by recovery, build capacity.

Chronic background stress, the persistent low-grade activation produced by constant digital stimulation, financial worry, professional uncertainty, and the feeling that you are always slightly behind, is a different animal entirely. It does not build capacity, it depletes and keeps stress elevated without the recovery that would allow recalibration.

The goal is not to eliminate stress. It is to become discerning about which stresses you choose to engage with, and to build recovery practices that allow your nervous system to return to baseline between them. One way to approach this is to treat episodes of acute or useful stress as experiments rather than as enemies. When you encounter a moment of real pressure this week, such as a high-stakes conversation or a demanding task, consciously frame it as data: notice how your mind and body respond, what strategies help, and what you learn about your own limits and capacities. By taking on this stance of curiosity, you transform stress from something to be avoided into information that can strengthen your adaptability.

In practice: At the end of each working day, identify one thing you were worried about that did not happen. This is not positive thinking. It is a nervous system recalibration and a deliberate practice of interrupting the brain's tendency to rehearse threat.

Try also tracking one "useful stress" episode this week. Observe it as it unfolds, then spend a few minutes debriefing afterwards: What did you notice? What helped you recover? What, if anything, will you do differently next time? Treating pressure as data encourages an adaptive rather than avoidant mindset, helping you become more resilient over time.

Examine the Beliefs You Inherited

The patterns we absorb in childhood regarding effort, security, and self-regulation become our unexamined operating system. My mother's belief in hard work built extraordinary things, but her views on worry, silence, and unspoken costs shaped both her and her children.

You do not have to reject the generation before you. The discipline, the belief in effort, and the commitment to independence are worth keeping. What is worth examining is whether you have also inherited the silent costs: the internalised stress, the inability to recover, the sense that rest is earned rather than required.

The brain changes in response to belief. What you believe about your capacity and future, especially regarding effort in an AI-driven world, shapes brain function.

In practice: Write down one belief you hold about your professional future in an AI world. Then ask: Does this belief support the brain you want to build? If not, what belief would?

Build Recovery Into the Structure of Your Day

Recovery is an active biological process, a window during which the nervous system recalibrates, the hippocampus consolidates learning, and the prefrontal cortex restores its capacity for complex judgement. Without recovery, performance degrades, not suddenly, but steadily, in ways that are often invisible.

Sleep is the most powerful recovery mechanism available to us. During deep sleep, the glymphatic system, the brain's waste-clearance network, flushes metabolic byproducts that accumulate during waking cognitive activity. Chronic persistent sleep deprivation leads to neurodegeneration and dramatically impairs self-regulation and decision-making skills that professional life demands. No amount of discipline, motivation, or caffeine compensates for chronic poor sleep.

Within the working day, deliberate pauses and genuine disengagement from digital stimulation, even briefly, allow the default mode network to activate. This is the neural network associated with perspective, creativity, and the synthesis of disparate information into coherent logic. The professional who never pauses is not more productive. They are less creative, less accurate in their judgements, and less capable of the kind of thinking that AI cannot reproduce.

Practice: Protect one twenty-minute period each day with no screen, no input, no agenda. Walk without a podcast. Sit without a task. This is the time during which your brain does its most integrative work.

The world my mother walked into before dawn was a different world from the one we now inhabit. The stresses were different, the tools were fewer, and the pace was slower. But the fundamental requirement has not changed: to sustain meaningful work across a life, you need a brain and a nervous system that can take the load, recover from the strain, and keep returning to full capacity.

In an age of artificial intelligence, where information is constant, and the pace of change is accelerating, this capacity is no longer a luxury but the foundation on which everything else depends. Starting where you are, moving your body regularly, becoming aware of your beliefs and thoughts, and deliberately building recovery into the structure of your days are all essential practices for protecting and strengthening your brain.

Your brain is always adapting, changing in response to what you repeatedly do, what you repeatedly think, and what you practise over time, and this is neuroplasticity, which matters more than ever because in an age defined by competition for your attention, you do not have the option to opt out.

CHAPTER THREE

The Attention War

How social media hijacks focus and how to take it back

I was sitting in a meeting room inside Parliament House when the parents placed their children's photographs on the table. They were ordinary images: school uniforms, birthday cakes, teenagers standing in sunlight with their whole lives ahead of them. Outside, political conversations continued as usual, but inside that room, we were facing consequences that could not be undone.

As the parents spoke, a pattern emerged that is now disturbingly familiar. It often begins with a single message that feels harmless or even kind, then shifts quickly into manipulation. An image is shared through an online platform, sometimes real and sometimes altered, and then used as leverage. A threat follows, money is demanded, and shame takes hold with such force that it overwhelms a young person's ability to think clearly or reach out for help. Sextortion is not a rare or unpredictable event. It is an expected outcome when developing brains, still building judgement and impulse control, are exposed to digital systems designed for speed, reach, and emotional impact rather than protection. These are not isolated cases.

What struck me most was not only the tragedy of what had happened, but the speed at which it unfolded. The adolescent brain is wired for belonging, and social rejection activates the same neural systems as physical pain. The capacity to pause, to put things in perspective, and to choose a response is still developing during these years. In that window, humiliation does not feel temporary or manageable. It feels overwhelming and absolute because, at a neurological level, it is experienced that way.

This is the attention war. These platforms are designed to capture and hold attention, to amplify emotion, and to drive engagement at scale. In doing so, they collide directly with the vulnerabilities of the developing brain. This is not about weakness or poor choices. It is biology meeting design, and the consequences are playing out in real lives every day.

Behind every feed is a nervous system. Behind every algorithm is a design choice. And behind the most extreme cases, the families I have sat with, the photographs they carry, are the consequence of treating both as someone else's problem.

This chapter is not primarily about those extreme cases. It is about the subtler, slower, more pervasive attention war that plays out in every brain, adult and adolescent, professional and student, that spends significant time inside algorithmically curated digital environments. It is about what that war is doing to our capacity for focused thought, emotional stability, and genuine connection. And it is about what we can do at the levels of individual behaviour, professional culture, and public policy to fight back.

The Architecture of Distraction

Attention is not simply what you happen to be thinking about. It is a limited cognitive resource and the bottleneck through which all learning, reasoning, creativity, and emotional regulation must pass. Whatever commands your attention commands, in a very real sense, your mind. This has always been true. What is new is the scale and the sophistication of the systems now competing for it.

Social media platforms are not accidentally compelling; they are the product of teams of engineers and behavioural scientists who understand human psychology and neural architecture with considerable precision and who have applied that understanding to the specific goal of maximising the time users spend on their platforms. They are the same mechanisms that neuroscientists study in addiction research, applied to digital environments and refined through billions of data points.

Infinite scroll eliminates the natural stopping cues that physical media provides. Autoplay removes the pause that lets a person decide to do something else. Notifications operate on intermittent-reinforcement schedules, the same mechanism that makes gambling compelling because unpredictable rewards elicit stronger approach behaviour than predictable ones. Like counts, share metrics, and comment notifications quantify social evaluation and make it public, triggering the same social-threat monitoring circuits discussed in the previous chapter.

Algorithms learn what you engage with, what you linger on, what you scroll past, what you click, and what makes your thumb stop moving, and they surface more of it. Not because they care what is true, or good for you, or likely to make you feel better about your life. Engagement is the metric, and attention is the product. You are not a customer of these platforms; we are what they sell.

The attention economy does not care about your well-being, your productivity, your sleep, or your children's development. It cares about time on the platform.

5.24 billion people social media users globally* (DataReportal, 2025)

We are now living in a deeply connected world, with 8.2 billion people on the planet and the majority, nearly 60 per cent, living in urban environments. Around 5.8 billion people use a mobile phone, most of them smartphones, and 5.6 billion are connected to the internet. Social media alone reaches 5.24 billion users, meaning almost two-thirds of humanity is now part of a shared digital ecosystem. While billions remain offline, the scale and speed of this global connectivity are unprecedented, shaping how we think, relate, and experience the world in ways that are only just beginning to be understood.

When nearly three-quarters of humanity participates in a system optimised for engagement rather than thriving, the effects extend far past individual screen-time habits. They shape the information environment in which we form beliefs, the way in which we assess the world, and the attention we can bring to every other area of our lives.

This means that when we find ourselves struggling to reclaim focus, it is not simply a test of individual willpower. We are up against networks of algorithms and financial incentives optimised to override self-regulation. Understanding this connection, from the neural circuitry of dopamine and reward to the economic systems invested in keeping us scrolling, clarifies why the attention war must be fought on multiple levels: biological, behavioural, and systemic.

What Distraction Does to the Brain

The brain is not built for constant task-switching. Executive attention, the capacity to hold a goal in mind, resist distraction, and sustain focus long enough for deep work, is one of the most metabolically expensive cognitive processes. It is also one of the most trainable. Like a muscle, it strengthens with use and weakens with disuse.

The cost of losing focus has been measured in what is often called the "23-minute rule," based on the work of attention researcher Gloria Mark and her colleagues. Their research shows that after an interruption, it takes an average of 23 minutes and 15 seconds to return to deep, focused work. In a typical knowledge-based role, where outcomes depend on sustained thought, even a few interruptions can erode an entire hour. If your attention is redirected just three times, most of that hour is spent recovering rather than producing.

The financial impact becomes clear when you apply this to real work. A lawyer or consultant billing $300 per hour effectively loses around $75 in value for every 15 minutes of disrupted focus. When this pattern repeats across days, teams, and organisations, the cumulative cost of distraction quickly grows to tens of thousands of dollars each year, alongside the hidden costs of reduced quality, errors, and mental fatigue.

When attention is repeatedly interrupted by notifications, by the reflexive check of the phone, by the awareness that there might be something new since you last looked, the neural circuits supporting sustained focus are not simply paused. They are pulled away from the task and partially redirected. Research on attention residue shows that even after returning to a task, a portion of cognitive resources remains tied to the interruption for minutes afterwards. The cost of a ten-second notification check is not ten seconds. It is closer to twenty-three minutes and 15 seconds of reduced cognitive capacity.

Over time, repeated interruption changes the brain's default expectations. A nervous system conditioned to expect stimulation every few minutes begins to experience the absence of stimulation as discomfort. Sustained focus feels harder, not because the task has changed, but because the brain has adapted to distraction. The threshold for boredom drops, and the tolerance for difficulty shrinks. The capacity for the deep, effortful thinking that produces genuinely original insight, what the psychologist Mihaly Csikszentmihalyi called flow, becomes increasingly inaccessible.

The work that AI cannot replicate, the detailed judgement, the complex clinical reasoning, the creative strategy that requires holding many variables simultaneously, is precisely the work that demands sustained and undivided attention. As AI absorbs routine cognitive tasks, the remaining distinctly human contribution becomes more concentrated in the capacities that divided attention most directly undermines.

For many professionals, attention is the central war of the working day. And most are losing it without realising the extent to which the battlefield has been designed against them.

The Anxiety Engine

Social media is not only an attention problem, it also amplifies anxiety in the following ways:

Fear of negative evaluation is one of the most common anxiety triggers. A post can be liked by three hundred people or ignored entirely. A comment can be celebrated or savaged, and performance is always visible, measurable and available for comparison. For people whose professional identity is closely tied to how they are perceived, which describes most of the audience for this book, this creates a chronic low-grade evaluation threat that the nervous system cannot simply get used to.

Doomscrolling, the compulsive consumption of negative news and alarming content, is another mechanism. Algorithms learn that threatening, fear-inducing content holds attention longer than neutral content. The feed becomes, gradually and then suddenly, a curated stream of crisis, outrage, and catastrophe, not because the world is only those things, but because those are the things that keep the thumb moving. The brain's threat-detection system, designed to scan for danger and respond, is activated repeatedly. And the user scrolls on, caught in a loop the platform has designed specifically to be self-sustaining.

> The most common mental illness in Australia:
>
> **Anxiety disorders**
>
> (3.4 million people, or 17% of the population, Australia Institute of Health and Welfare)

Mental overload refers to the volume of information that passes through a typical social media feed in an hour, exceeding what a person in any previous era would have encountered in a month. The brain is not, and was never, equipped to process this volume of rapidly shifting content without cost. Working memory becomes saturated and the capacity for reflective, deliberate thinking is displaced by reactive, automatic processing. We become more susceptible to misinformation, more prone to emotional rather than reasoned responses, and less capable of the kind of careful attention that professional and personal life actually require.

These mechanisms do not require heavy use to have an effect. The research shows measurable impacts on anxiety and attention at levels of social media consumption that most users would consider moderate. The dose at which these systems become problematic is lower than most people assume, and lower than the platforms have any incentive to acknowledge.

The Adolescent Brain: A Specific Vulnerability

Everything described above applies to adults. For adolescents, the picture is significantly more serious, and the stakes are categorically different.

Adolescence is the second critical period of rapid brain development, after early childhood. The prefrontal cortex, which governs impulse control, long-term planning, and emotional regulation, is still being constructed, a process that is not complete until the mid-twenties. At the same time, the reward and social-evaluation systems are running at heightened sensitivity. Adolescents are neurologically primed to care intensely about their peers, to seek novelty and excitement, and to weigh immediate social feedback heavily against long-term consequences.

This is not a design flaw but an important adaptive phase, one that serves important developmental purposes. The problem is that social media platforms are not safe environments for development. They are engagement-optimised systems that have learned, with extraordinary precision, how to navigate adolescence.

We do not allow children to drive until we judge their regulatory systems mature enough to handle the responsibility. We have not applied the same reasoning to social media, and the evidence of that failure is visible in clinical waiting lists and school counsellor caseloads.

Research on adolescent brain development and the wider scientific literature consistently support the same conclusion: the minimum age at which unregulated social media access should be permitted is higher than most platforms have acknowledged, and the type of content that algorithmic systems serve to young users requires reform.

Reclaiming Attention: What Actually Works

The attention war is winnable, not by withdrawing from digital life, which is neither realistic nor, given its genuine benefits, desirable, but by engaging with it on different terms. The research on how to regain attention is actionable.

Design your environment before your environment designs you.

The most durable changes are environmental. Every barrier you place between yourself and reflexive social media use, moving apps off the home screen, requiring an extra step to access them, removing them from the bedroom entirely, reduces use without requiring ongoing self-discipline. You are not relying on deciding, in the moment, to make the better choice. You are changing the environment, so the better choice is easier.

I moved my social media apps off my home screen after a period in which the algorithm surfaced content I had not sought and could not ignore. My nervous system responded to those images in ways entirely predictable given what we know about social comparison and threat detection. My sleep and mood deteriorated, and my attention was fragmented. The effect was not dramatic; it was quiet and cumulative, which is exactly how environmental change to neural architecture works. The brain adapts to what it is repeatedly exposed to. If we change the exposure, the brain will adapt.

In practice, try removing social media apps from your phone's home screen. Require yourself to search for them to open them. This single increase in friction reduces use without requiring any further decision-making. It is the lowest-effort, highest-return attention intervention supported by the research.

Protect deep work with structural defence, not willpower.

A knowledge worker operating in a state of chronic partial attention, never fully on the task, never fully off, can design specific windows for focused work, two hours minimum, and make them protected: phone on silent, notifications disabled, door closed, or status set to unavailable. Do not rely on colleagues' goodwill or your own resistance to distraction. Build environments that make distraction physically harder.

In practice: Block one 2-hour window daily as non-negotiable deep work time. No meetings, no notifications, no social media, and treat it with the same firmness you would a client appointment. It will feel uncomfortable at first; that discomfort is the attention system recalibrating from distraction to focus.

Use social media with purpose, not as a default.

The most damaging pattern in social media use is not volume; it is unconsciousness. The absent-minded scroll that fills every moment. The reflexive check that happens whenever you feel mildly bored, anxious, or uncertain. The phone that appears in your hand without a conscious decision to reach for it.

This pattern is damaging not primarily because of what the content does, though that matters, but because of what the behaviour itself trains. It teaches the brain that external stimulation is the default response to internal discomfort. Every time you reach for your phone when you feel slightly unsettled, you reinforce the pathway that makes it the automatic response to that feeling. Over time, the tolerance for internal discomfort states, boredom, uncertainty, creative friction, and the productive discomfort of difficult thinking diminishes.

Those internal states are not problems to be solved. They are the conditions in which some of the most important cognitive work happens. The moment before a solution arrives and the pause in which a question becomes clearer. The discomfort means you are thinking about something genuinely hard.

In practice: When you reach for your phone, pause for three seconds and ask: Did I decide to do this, or did my hand just move? If the latter, put it down. You do not need to avoid your phone. You need to make using it a choice rather than a reflex.

Distinguish the platform from the connection.

Social media use is not monolithic or passive consumption; activities such as doomscrolling, reading, watching, and comparing consistently show stronger associations with anxiety and reduced well-being than active connection, direct messaging, meaningful comment exchange, and coordinating face-to-face meetings. The platform is a tool, and the question is what you are using it for.

If you use social media primarily to stay in contact with specific people you care about, and those contacts reinforce rather than substitute for real-world connections, the evidence suggests the effects are considerably more benign than if you use it primarily to scroll a feed of curated content from people you have never met. The goal is to use social media in ways that serve the connections and purposes that matter to you, rather than the engagement metrics that matter to the platform.

In practice: Before you open a social platform, identify one specific person you intend to connect with or one specific piece of information you intend to find. If you cannot name one, close it. Purpose transforms the neurological impact of use.

The War Worth Fighting

The attention war cannot be won at the individual level alone. Personal habits matter, and they are the right place to start because they are the place where you have the most immediate agency. But the systems we are contending with are vast, well-resourced, and specifically designed to resist the kind of friction-creation and conscious use that individuals are advised to practise.

This is why the policy work matters. The Australian legislation on minimum age requirements was imperfect and insufficient, and all legislation in a fast-moving space is, but it was not meaningless. It asserted that democratic societies could make collective decisions about the digital environments in which their citizens, and especially their children, develop. If the technology that created the attention war can be redirected to detect and support the people most harmed by it, we are not simply mitigating damage. We are demonstrating that a different version of the relationship between technology and human neurobiology is possible, and one in which design serves development rather than exploiting it.

Thriving minds are not built by accident. They are built on purpose. In a world where systems profit from distraction, being intentional has never been more important or more possible than it is right now. Try to put your phone down to protect your focus and then help us advocate for more support to build something different. The war is real. So is our capacity to fight it.

CHAPTER FOUR

Your Brain Under Pressure

What uncertainty, speed and overload are doing to us

I remember the exact moment I stopped thinking clearly.

It was not a dramatic collapse. There was no single event I could point to afterwards and say, "That is where it broke." It was more like a tide going out so gradually that I did not notice the waterline dropping until I was standing on dry sand, wondering what had happened to the ocean. I ran a research lab in San Francisco with 17 people working for me. I was publishing, presenting, applying for grants, raising children, maintaining the appearance of a life that, from the outside, seemed to be running exactly as it should. And somewhere in the middle of all of that, I lost access to the part of my mind that could think with any depth at all.

I would sit at my desk with a paper I needed to read and read the same paragraph four times without retaining it. I would start a grant application and, twenty minutes later, find myself rewriting the opening sentence repeatedly without making any progress. In meetings, I could speak the language of neuroscience fluently, but I was aware, in a way I had not been before, that I was assembling words rather than thinking. I was a neuroscientist, and I did not recognise, until much later, that what I was experiencing was my own brain under sustained, unrelenting pressure, doing exactly what brains do when pressure exceeds recovery for long enough.

It was shutting down the expensive functions to keep the essential ones running.

The brain under chronic pressure does not fail dramatically. It fails gradually, in the capacities you value most and it does so quietly enough that you can mistake the decline for a personal failing rather than a physiological response.

This chapter is about what pressure does to the brain, and specifically the kind of pressure that characterises professional life in the AI era: the uncertainty of a landscape that keeps changing, the speed of information that keeps accelerating, and the cognitive load of a world that never turns off. It is about the neuroscience of how the stress response works, what happens when it cannot return to baseline, and what the condition we call burnout is at the level of the body, brain and mind. And it is about what the research shows actually helps, evidence-based interventions that allow a pressured nervous system to recover its capacity without requiring you to abandon the work that gives your life meaning.

The Brain Is Not Built for This

The human stress response is one of the most elegant and precisely calibrated systems in biology. Faced with a genuine threat such as a predator, a physical danger, or an acute social crisis, the hypothalamic-pituitary-adrenal axis activates within seconds. Cortisol and adrenaline flood the system, and the heart rate increases. Blood is redirected from the digestive system and the prefrontal cortex toward the muscles and the sensory organs. Attention narrows to the immediate threat, and the prioritised memory is consolidated; as a result, complex reasoning is suspended.

This is not a malfunction but a masterful design, where the presence of acute danger is the last thing you need to worry about. You need speed, strength, and reactive precision, and the stress response delivers all three.

The problem is that this system evolved for a world of acute, bounded threats, a predator that either catches you or does not, a storm that either passes or does not, a confrontation that either resolves or does not. It was not designed for the chronic and unresolvable pressures that characterise modern professional life: the inbox that refills as fast as you empty it, the performance metrics that never stabilise, the professional landscape that keeps shifting before you have finished adapting to the last shift, and most acutely now, the uncertainty of working in a world where the tools, the roles, and the definitions of expertise are all in motion simultaneously.

When the stress response activates repeatedly without resolution and when the nervous system never returns to baseline, when the perceived threat is not a lion that runs away but a professional environment that simply continues, the consequences compound in ways that are specific, measurable, and in many cases, reversible if caught early enough.

Uncertainty Is the Hardest Stress

Not all stressors are equal. The research on stress neuroscience consistently finds that uncertainty, the condition of not knowing what will happen, when, or what it will require of you, is neurologically more costly than most other forms of pressure, including pressure that is severe but predictable.

A landmark study by de Berker and colleagues at University College London had participants turn over rocks on a screen; beneath some rocks was a snake, and if there was one, the participant received a small electric shock. The researchers varied the probability that any given rock contained a snake from 0% to 100% and measured stress responses across the range. The finding was striking: peak stress did not occur at 100% probability, where a shock was certain to occur.

It occurred at approximately 50% probability, at maximum uncertainty, and where the participant simply did not know.

Certain bad outcomes are, neurologically, easier to bear than uncertain ones. When you know what is coming, the brain can prepare and allocate resources to respond. When you do not know, you must maintain continuous vigilance, such as scanning, anticipating, and running threat simulations, which is metabolically expensive and physiologically exhausting.

The most stressful professional environment is not the one with the heaviest workload. It is the one where the rules keep changing, the criteria shift without notice, and the future remains persistently opaque.

This is a precise description of professional life in the age of AI. The AI transformation of knowledge work has created the conditions for maximum neurological stress: capability boundaries are shifting, role definitions are being renegotiated, the value of specific expertise is uncertain, and the pace of change ensures that as one uncertainty resolves, another opens. The professional who felt secure in their skills eighteen months ago may feel considerably less so today, and not because their skills have diminished, but because the landscape in which those skills operate has shifted enough to make their value feel less certain.

Survey findings indicate that a significant majority of professionals report an accelerating pace of change in their workplace (Australian HR Institute, 2024)

Understanding this mechanism does not eliminate the uncertainty. But it does something almost as valuable: it converts the experience from a personal failing of why I cannot just cope with this, into a physiological response. You are not struggling because you are weak or underprepared. You are struggling because your nervous system is doing exactly what nervous systems do in conditions of sustained uncertainty. That reframe matters because the pathway out of helplessness runs through understanding, and it is much harder to find when you are blaming yourself for the symptoms of a normal biological response.

What Information Overload Does to the Thinking Brain

Alongside uncertainty, the sheer volume of information that characterises modern professional life imposes a specific and well-documented cognitive cost.

Working memory, the brain's active processing workspace, where information is held and manipulated in real time, holds approximately four to seven chunks of information at once. It maintains them for roughly twenty seconds before they begin to decay unless actively rehearsed.

It is, as the cognitive scientist George Miller described it in a paper that remains one of the most cited in psychology, 'the magical number seven, plus or minus two.'

The information environment we now inhabit was not designed with this constraint in mind. It was designed around engagement metrics. The volume of content arriving through email, messaging platforms, news feeds, professional networks, and AI-generated outputs on a typical working day exceeds, by several orders of magnitude, what working memory can process without cost. The result is not simply that some information gets lost, though it does. The result is that processing high volumes of information degrades overall processing quality.

When working memory is saturated, the brain defaults to heuristic thinking, such as pattern-matching shortcuts that allow fast responses with minimal cognitive effort. This is adaptive when speed matters more than accuracy. It is significantly less adaptive when the task requires judgement, careful weighing of evidence, or the kind of original synthesis that distinguishes expert human thinking from competent algorithmic output. Overloaded working memory produces faster, shallower, more error-prone cognition. It is the neurological explanation for why your best thinking rarely happens on your busiest days.

Every email you answer while holding a complex problem in mind is not simply a distraction. It is a partial erasure because working memory is a finite slate, and writing on it costs space that would otherwise hold something else.

The AI era has added a specific new dimension to this problem. The availability of AI tools that can rapidly generate summaries, analyses, and first drafts has created a professional environment in which the volume of content to be reviewed, approved, contextualised, and acted upon has increased substantially, even as the time available for each item has shrunk. The cognitive demands of AI-assisted workflows, such as prompting, evaluating, editing, and integrating, are non-trivial and sit atop existing demands rather than replacing them.

For many professionals, the experience of AI augmentation in the first year of adoption is not relief. It is a new kind of overload: more output to review, more decisions about what to use and what to discard, more context-switching between their own thinking and the machines, and the added metacognitive load of evaluating whether the AI's confident-sounding output is actually correct.

The Compound Effect: How Pressure Builds

What makes sustained professional pressure particularly insidious is that its effects are not linear; they compound, and a single high-pressure week can lead to fatigue and temporary cognitive decline. Sleep, exercise, and a quieter weekend restore most of the lost capacity. This is the normal stress-and-recovery cycle that a healthy nervous system manages without lasting consequence. The problem begins when the recovery window is consistently insufficient and when the next high-pressure period begins before the previous one has fully resolved, when sleep is routinely shortened, when weekends offer stimulation rather than rest, and when the perception of being perpetually behind becomes the baseline rather than the exception.

The experience of this progression, from the inside, is one of increasing fragility. Situations that would once have been manageable provoke disproportionate reactions. Small setbacks feel significant. The capacity for perspective, for holding current difficulty against a longer time horizon and finding it survivable, diminishes as irritability increases and patience shortens. The professional who was once known for their equanimity under pressure now responds to minor frustrations with a strength that surprises them.

This is not a personality change. It is a neural architecture change that, at this stage, is still reversible but will continue to progress if the conditions driving it remain unchanged.

> *1 in 2 Australians Facing Workplace Burnout - Young Adults and Caregivers Leading the Trend (Beyond Blue, 2025)*

Burnout is now a defining feature of modern working life in Australia, with a community survey commissioned by Beyond Blue finding that around half of workers experienced burnout in the past year, and that young people and parents are at the highest risk. Despite being early in their careers, those aged 18–29 report the highest levels, driven by heavy workloads, limited support, and inflexible working conditions, all layered atop broader pressures such as housing costs, financial strain, climate anxiety, and loneliness. Small business owners appear particularly vulnerable, reflecting the relentless demands and blurred boundaries of their roles. Together, these findings suggest burnout is not simply an individual failure to cope, but a systemic signal that the way we work and the conditions we expect people to endure need an urgent redesign.

The trajectory from sustained pressure to burnout is not a cliff. It is a gradual slope, and most people do not recognise where they are on it until they are considerably further down than they realised. The warning signs are increased cynicism, fast reactions, anger, reduced engagement, a sense of professional ineffectiveness that persists regardless of objective performance and are often misread as motivational problems or character failings rather than as the physiological indicators they actually are.

What Burnout Actually Is

Burnout has been in the clinical and management literature for decades, but it has also been significantly misunderstood, treated as a synonym for exhaustion, a state of being very tired, something that a holiday or a long weekend might address. Neuroscience tells a more specific and more serious story. Burnout, as characterised by Christina Maslach's foundational research and subsequently validated across clinical populations, has three distinct dimensions: emotional exhaustion, depersonalisation, and reduced sense of personal accomplishment.

Emotional exhaustion is the feeling of having nothing left to give. It draws on resources that are replenished primarily through sleep and rest. When those resources are chronically depleted faster than they are replenished, the capacity for emotional regulation degrades, and experiences that were once manageable become unbearable.

Depersonalisation is the numbing, the withdrawal of empathy, the shift toward treating colleagues and clients as problems rather than people, and, neurologically, a protective response. When the system has no further capacity for emotional engagement, it creates distance as a form of self-preservation. This is not a moral failure; it is the nervous system doing what nervous systems do when no other resource is available: it reduces the cost of engagement by reducing engagement itself.

The reduced sense of personal accomplishment and the loss of any felt sense that one's efforts matter, even when objective evidence suggests they do. The neural circuits that connect effort to satisfaction, that produce the internal signal that something has been achieved and is worth continuing, become uncoupled from the outcomes they would normally register. Work that was once intrinsically motivating becomes hollow and the professional finds themselves unable to feel what they intellectually know they should feel.

Burnout is not a motivational failure. It is a neurological state in which the systems that make work meaningful, manageable, and human have been depleted beyond their capacity for self-restoration.

The distinction matters for two reasons. First, because it changes what recovery requires. If burnout is just exhaustion, a holiday is probably sufficient. If it is a structural change in neural architecture, then recovery requires the specific biological inputs that rebuild those structures: sustained sleep, regular movement, genuine social connection, and, critically, a sustained reduction in the chronic stressors that drove the degradation. A two-week holiday does not rebuild hippocampal volume, whereas months of changed conditions do.

Second, because it changes how we should talk about it. A professional who is burning out is not failing at their job. They are experiencing a medical condition with identifiable neurological correlates, driven by environmental conditions that in many cases are not within their individual control to change. The language of personal resilience, as well as 'find your grit' and 'build your mental toughness,' applied to burnout, is not only unhelpful but also physiologically wrong, and it delays the recognition and structural response that recovery actually requires.

The AI-Era Pressure Profile

Every era of rapid technological change has produced professional stress. The industrial revolution, the computerisation of knowledge work, and the arrival of the internet each disrupted established roles, required the acquisition of new skills, and generated the uncertainty and overload that the nervous system finds most costly.

What distinguishes the current moment is not simply the pace of change, but the nature of what is changing.

Previous technological development led to physical labour, then to routine clerical work, and then to structured data processing. The skills at risk were generally distinct from those that distinguished professional expertise: a lawyer's ability to reason about precedent, a doctor's capacity for differential diagnosis, a teacher's skill in reading a classroom and adapting in real time. Those capacities felt safe precisely because they were complex, contextual, and apparently irreducible to rule-following.

Generative AI disrupts at a different level. It can produce legal arguments, clinical summaries, lesson plans, research syntheses, and planned analyses that, in many cases, are useful and, in some, indistinguishable from competent professional output. This does not mean AI is replacing professional expertise, and the argument throughout this book is that the distinctively human capacities of judgement, social connection, ethical reasoning, and contextual wisdom remain continue to be irreplaceable does mean that the anxiety about what those capacities are, whether they are sufficient, and how to demonstrate their value in a landscape that has shifted, is entirely rational.

The professional under AI-era pressure is not simply dealing with a heavy workload or a difficult period. They are dealing with an identity question: What am I for? And identity questions, as the social neuroscience literature abundantly shows, activate the threat response with particular intensity. We are social animals whose sense of safety is deeply tied to our sense of place. When that sense of place becomes uncertain, the nervous system does not treat it as an interesting intellectual problem but as danger.

The anxiety professionals feel about AI is not about the technology itself. It is about identity and whether the capacities that define who you are professionally will remain valuable tomorrow. That is among the most neurologically activating uncertainties a human being can face.

Understanding this reframes the appropriate response. The answer to AI-era pressure is not, primarily, to learn more tools or acquire more technical competence, though both have their place. It is to stabilise the nervous system under pressure, so that the thinking required to navigate the transitions is available. A brain in chronic threat-response mode cannot do the expansive, creative, collaborative thinking that the transition demands.

Working Under Pressure Without Fracturing

The neuroscience of stress is, in one respect, a deeply hopeful literature. The brain is plastic, and the structures that chronic pressure degrades, such as the hippocampus, the prefrontal cortex's regulatory capacity, and the dopaminergic reward circuits, are the same structures that respond to recovery. Given the right conditions, they rebuild. The question is: what are those conditions?

What follows is not a stress-management checklist. It is a summary of interventions with the strongest evidence base for restoring and protecting brain function under sustained professional pressure, organised by the specific mechanism each addresses.

For the threat response: name it to tame it

One of the most robust findings in affective neuroscience is that labelling an emotional state reduces its intensity. When you name what you are feeling, such as not 'I am stressed' generically, but 'I am feeling genuine fear about whether my skills will still be valued in two years', the prefrontal cortex engages more directly with the experience, and the stress reactivity measurably decreases. This is the neural mechanism underlying many forms of psychotherapy, and it can be applied in any clinical setting.

The practice of naming, which converts a diffuse, activating sense of threat into a specific, articulated concern, serves two functions. It reduces the physiological cost of the experience in the moment. And it brings the concern into the domain of solvable problems, or at least assessable ones, rather than leaving it as ambient dread, which is expensive. Whereas a specific concern can be examined, contextualised, and responded to.

In practice: At the end of each working day, write one sentence naming the specific uncertainty that most activated your threat response. Not 'I'm worried about everything' but one specific thing. Naming reduces activation, which you can begin to regulate.

To reduce cognitive overload, protect the processing window.

Working memory has a fixed capacity, but the conditions under which it operates can be protected or degraded by the environment you create around the tasks that require it.

The most evidence-supported intervention for cognitive overload is not time management. It is task isolation and the deliberate practice of engaging with one cognitively demanding task at a time, in an environment specifically designed to protect the processing required for that task. This means no open email, no background notifications, no ambient conversations, and no partially completed other tasks visible on the screen. The brain cannot multitask at the level of complex cognition. What it does instead is rapid serial switching, which is slower, more error-prone, and more fatiguing than sustained focus on a single task.

For professionals using AI tools: the cognitive effort AI-assisted workflows require is real and should be managed accordingly. If you are using AI to generate a first draft, review it in a separate, focused window from the one in which you are doing your own thinking. Do not attempt to simultaneously prompt, evaluate, and think; these are three distinct cognitive modes, and running them concurrently is expensive. Try sequencing them instead.

In practice: Before beginning any cognitively demanding task, spend 90 seconds clearing your working memory. Take a dedicated "holding pen" notebook or notepad—something you keep close to your workspace and use only for this purpose. Write down the three most pressing other things your mind is holding, parking them in that notebook. By consistently placing thoughts in the same physical spot, your brain learns there is a routine, reliable place for temporary worries, reducing friction and making the ritual automatic. Once those lingering items are on paper and parked, your brain can let them go. This simple act of externalising and giving your thoughts a concrete home consistently improves the quality of subsequent focused work.

For compounding stress: build hard stops into the structure of the day

The compounding of stress across days, weeks, and months occurs primarily because recovery is treated as optional and occurs only when there is time for it, which, in a demanding professional environment, is rare. The research is unambiguous: recovery is not optional and is the mechanism through which the nervous system maintains the capacity to sustain performance. Without it, capacity degrades; with it, it can be maintained and, under the right conditions, built.

Hard stops mean specific times at which work ends, and the nervous system begins its recovery, and they are not a concession to weakness. They are a structural feature of sustainable, high-performance and the professional who works until they are finished rarely finishes in a way that allows genuine recovery. The professional who stops at a specific time, regardless of whether everything is done, protects the neurological capacity required for tomorrow's work.

This is particularly important in an era of AI-augmented work, where the boundary between working and not-working has become genuinely unclear. If your AI tools are running in the background, generating outputs you might check at any moment, the nervous system does not experience that as rest. It experiences it as low-grade vigilance, which is physiologically indistinguishable from low-grade stress.

In practice, choose a stopping time and a calendar blocking, device downtime, and AI tools being closed. Begin this not with the intention of catching up tomorrow, but with the understanding that what you are protecting in those recovery hours is the neurological capacity without which tomorrow's work will be worse than today's.

For burnout: address the environment, not only the individual

The most important thing to understand about burnout recovery is that it cannot happen in the environment that produced the burnout. If the conditions, such as volume, uncertainty, and a sense of ineffectiveness, do not change, then the interventions applied to the individual are working against a current too strong to swim against indefinitely.

This means that burnout recovery is partly an individual responsibility and partly an organisational and structural one. Sleep hygiene, exercise, social connection, and deliberate recovery practices are all evidence-based and worth implementing. They are also insufficient in isolation if the environment continues to deplete at a rate that exceeds what any individual intervention can restore.

For leaders and managers: the organisational conditions that produce burnout, such as chronic overload, unclear expectations, lack of autonomy, lack of recognition, and perceived unfairness, are well-documented, reliably predictive, and largely within your control. Creating environments in which people can do their best work without systematically destroying the capacity that makes it possible is not a welfare initiative but a performance strategy.

In practice: If you are experiencing the symptoms of burnout, such as emotional exhaustion, cynicism, and a persistent sense that one's efforts do not seem to matter, the first and most important step is to name it for what it is: a brain state produced by conditions, not a personal failing. Then identify one condition that is within your power to change. Not all of them, but one, as a change in brain states begins with the smallest available lever.

For the uncertainty of the AI era: anchor to what does not change.

In the uncertainty of the AI era, the most reliable anchor is what remains constant. When the external environment is changing rapidly, stability comes from a clear understanding of your core values, strengths, and commitments. The qualities that AI cannot replicate, including relationships, ethical judgement, wisdom, and the capacity to be fully present with another person, do not lose their importance as technology advances. Grounding your professional identity in these enduring human capabilities, rather than in technical skills that may evolve or be automated, provides both protection and clarity. Professionals who understand what they stand for and the uniquely human value they bring are better positioned to navigate change with confidence rather than fear.

In practice: Write down three things you do professionally that require your specific human judgement, presence, or experience and things that a capable AI could assist with but not replace. Read them when the uncertainty feels loudest. You are not defending yourself against change and are orienting yourself within it.

The Brain That Survives Pressure

I came back from that period in San Francisco, not all at once, and not without cost. There are things from those years that I think I will always carry in some form. But I came back to the capacity for deep thinking, for general engagement with complex ideas, for the kind of pleasure that had attracted me to neuroscience in the first place.

What brought me back was not a heroic effort but the complete opposite. It was, eventually, reducing the load enough that my nervous system could begin to recover, and then systematically building the conditions that research showed would support that recovery: movement I treated as medicine, connection I stopped outsourcing to screens, and work I allowed to end at a specific time.

I also had to make peace with something that took me longer to accept than it should have, that the brain under unsustainable pressure does not perform sustainably. The body keeps the score, as the trauma researcher Bessel van der Kolk has written, and the brain and mind, which are part of the body, keep it precisely.

Pressure is not the enemy, but pressure without recovery is. The brain that learns to move between activation and restoration that treats recovery not as a reward for completed work but as the mechanism that makes work possible, and is the brain built to last.

The world we are working in is not going to become less pressured. The pace of change will continue, and the volume of information will keep increasing. The uncertainty of the AI transition will persist for years, perhaps decades, as the implications continue to unfold across every profession and institution.

The question is not whether to find a less pressured world. It is whether to build a brain that can move through this one without fracturing that can sustain the effort, tolerate the uncertainty, recover from the overload, and remain capable of the distinctively human work that pressure, if we let it, can either destroy or deepen.

That brain is available to you. It is built the same way every important thing is built deliberately, incrementally, over time, with the understanding that what you repeatedly do is what you become.

The pressure is real. So is the capacity to bear it, and to remain, under it, fully yourself.

CHAPTER FIVE

The Human Upgrade

Food. Sleep. Movement. Energy

There is a particular kind of irony in spending your career studying the brain while systematically neglecting the body that keeps it alive.

I did this for years. I understood, at a research level, that the brain is not a floating consciousness but a metabolically expensive organ that consumes about 20% of the body's energy, is highly sensitive to inflammation and fluctuations in blood glucose and rebuilds itself during sleep and movement. Yet I would work through lunch, sleep five hours, and spend my afternoons in a cognitive fog, blaming my career rather than the demands I was placing on my biology.

It took a burnout described in the previous chapter and the gradual loss of deep thought, the sense of assembling words rather than thinking them, to bring me back to a truth that should have been obvious: you cannot separate the quality of your life from the quality of your biology. Your life is not a programme running on neutral hardware; it is tissue, blood flow, electrical activity, and chemical signalling, shaped by what you eat, how you move, how long you sleep, and how well you restore the energy you spend.

This chapter is about the four biological fundamentals that research shows are the most powerful, accessible, and consistently underused levers for building a brain that thrives under pressure. We will begin with movement, because it is the most underestimated of the four, and because understanding it properly reframes everything that follows.

Movement: The Most Underrated Health Tool You Have

When most people think about exercise, they think about the body, cardiovascular fitness, muscles or weight management. These are real benefits, and they matter; however, they are not the only important thing that the movement does.

Movement is the most powerful direct intervention for health, yet most professionals either do not use it or use it in ways that fail to capture its full neurological benefits. I spoke with Associate Professor Tara Walker from the Queensland Brain Institute on the Thriving Minds podcast about her research on exercise and neuroplasticity, and what she described changed my entire thinking about movement.

Exercise begins in the muscle, and when the skeletal muscle contracts, not gently, but with genuine effort, it releases a cascade of signalling molecules into the bloodstream that researchers have only recently begun to characterise fully. These molecules, called exerkines, are signals evolved to communicate between the body's working tissues and the organs that support them, including, critically, the brain.

The region most responsive to exercise-induced brain-derived neurotrophic factor (BDNF) is the hippocampus, the structure central to memory consolidation, spatial reasoning, and the regulation of the stress response. BDNF is sometimes described as a fertiliser for the brain, that is, a growth protein that physically changes the structure of neural tissue, and exercise is one of the most reliable ways to increase it.

The hippocampus is also among the structures most vulnerable to chronic stress. Exercise counteracts stress's negative effects. Multiple well-controlled studies have now demonstrated measurable increases in hippocampal volume in adults who engage in sustained aerobic exercise programs, with these increases corresponding to improvements in memory function and stress regulation.

This is not something that happens only in young, healthy and older adult research participants. happens in people with depression and anxiety disorders, where exercise-induced new brain cells, also called neurogenesis in the hippocampus, is now understood as one of the mechanisms underlying the antidepressant effect of physical activity and in people who have been sedentary for years and begin moving with modest, consistent effort.

The inactivity spiral and how to break it

Understanding why people do not move more requires understanding that inactivity is self-reinforcing at the level of dopamine. Dopamine, the neurotransmitter most associated with motivation and goal-directed behaviour, downregulates with inactivity. The less you move, the less dopamine signalling is available to motivate movement. The less motivated you feel, the less you move.

The solution to this cycle is not intensity, but simply getting started. Research on exercise initiation shows the barrier is not the activity itself but the move from rest to action. Once movement begins and the body has been in motion for ten minutes, the neurochemical environment shifts, and motivation rises. The resistance that felt insurmountable recedes.

Ten minutes is enough to begin the cascade. Not ten minutes of high-intensity effort, but ten minutes of movement, a walk, a gentle cycle, anything that gets the muscles contracting and blood moving. The probability that you will continue moving, today and tomorrow, increases.

150 minutes of moderate physical activity per week,

World Health Organisation 2020 guidelines

Strength training deserves a separate mention, because it is consistently undervalued in discussions of brain health. Resistance exercise improves insulin sensitivity, is relevant to glucose regulation and cognitive strength (as we will discuss in the food section). It is specifically associated with improvements in executive function: planning, cognitive flexibility, and the capacity to inhibit impulsive responses. The mechanism includes BDNF and the reduction of inflammatory markers that impair prefrontal network function. For professionals negotiating the complex, variable decisions that AI-era work increasingly demands, the brain health benefits of strength training are directly relevant to performance, not just health.

In practice: Commit to one form of movement you will actually do, three to five times per week. A walk you do consistently for a year does more for your hippocampus than a gym programme you abandon after six weeks. Start with ten minutes if thirty feels impossible.

Awe: Nature and the Healing Brain

How is it that time in nature, whether in a garden, a park, or out in the ocean, can shift how we feel so quickly? Why does music, movement, or being part of something bigger than ourselves leave us calmer, clearer, and more connected? Why do practices that create meaning, whether spiritual, cultural, or therapeutic, so often improve both mental and physical health? Research led by Dacher Keltner at the University of California, Berkeley, points to a common mechanism: awe.

Awe is a state that arises when we encounter something beyond our usual way of thinking. In those moments, the brain has to update and reorganise what it knows to make sense of what it is experiencing. It changes our physiology, our attention, and our sense of self.

This is a tool we can apply to the human upgrade. These experiences pull us out of narrow, stressed, self-focused thinking and shift us toward connection, perspective, and regulation. They quiet the noise, reduce stress, and open up new patterns of thought and response. This is neuroplasticity in real time, driven not by effort alone, but by experiences that expand the brain beyond its current limits. Across cultures and in everyday life, awe is neither rare nor reserved for extraordinary moments. Research shows that people experience it in simple and powerful ways, through witnessing courage and kindness in others, spending time in nature, moving together in dance or shared rituals, listening to music, engaging with art, or reflecting in spiritual practice. It can arise in quiet moments of insight and in the most profound experiences of life, including birth and death. They are different pathways into the same state, one that shifts the brain, broadens perspective, and supports the conditions for growth and healing.

Awe is not just a pleasant feeling or a passing moment. Whether it is studied in a laboratory or experienced in real life, out in nature or in connection with others, awe has measurable effects on the brain and body. It shifts our physiology, softens the focus on self, increases our capacity to connect with others, strengthens our sense of belonging, and brings a deeper sense of meaning. Each of these changes supports mental and physical health.

In practice: Step outside for a few minutes, look up at the sky, notice something in nature, or pause to take a slow breath, bringing your attention back to your body.

Food as Medicine

The brain accounts for roughly 20% of the body's resting energy expenditure. It is, calorie for calorie, the most metabolically demanding organ in the body, and it is almost entirely dependent on a continuous, stable supply of glucose for its energy. When that supply is stable, executive function operates effectively; when it is volatile, spiking after high-sugar meals and crashing several hours later, cognitive performance degrades in ways that are measurable, predictable, and largely preventable.

This is not a subtle effect, and research on glycaemic variability and cognitive performance consistently shows that the hours following a high-glycaemic meal, such as white bread, refined cereals, sugary drinks, and ultra-processed foods, are associated with reduced working memory capacity, impaired attention, and increased emotional reactivity. The blood sugar crash that follows a spike does not feel like low blood sugar. It feels like difficulty concentrating, mild irritability, a craving for something sweet, and a sense of brain fog that is easy to attribute to the demands of the day rather than to what you ate three hours ago.

Over time, repeated glycaemic volatility contributes to chronic low-grade inflammation and elevated circulating inflammatory markers that cross the blood-brain barrier, impairing the function of prefrontal networks that govern attention, impulse control, and complex reasoning. This is the neurological explanation for a phenomenon that most professionals describe as brain fog: the persistent, difficult-to-source sense that thinking is harder than it should be, that concentration requires more effort than it once did, that the clarity that used to characterise good working days has become less reliable.

Brain fog is often described as a psychological experience. In many cases, it is a metabolic one and the direct neurological consequence of an inflammatory dietary pattern meeting a sleep-deprived, chronically pressured nervous system.

The gut–brain axis system, most professionals have never heard of

One of the most significant developments in neuroscience over the past two decades has been the characterisation of the gut-brain connection, the communication network between the gastrointestinal system and the central nervous system, mediated by the vagus nerve, the immune system, and a remarkably complex ecosystem of approximately thirty-eight trillion microorganisms that inhabit the human gut.

The gut microbiome is an active participant in brain function. Gut bacteria produce neurotransmitter precursors, including approximately ninety per cent of the body's serotonin, as well as short-chain fatty acids that influence inflammation, the integrity of the blood-brain barrier, and the function of glial cells and the support cells of the brain's neural network. The composition of the microbiome and which bacterial species are present, in what proportions, and in what diversity is directly influenced by diet and is directly associated with mental health outcomes.

Diets high in ultra-processed foods, refined sugars, and seed oils reduce microbial diversity and promote the growth of inflammatory bacterial species. Diets high in fibre, fermented foods, and a diverse range of plant matter are associated with lower rates of depression, anxiety, and cognitive decline. The research group led by Professor Felice Jacka at Deakin University has been particularly influential in establishing these associations with rigorous clinical data, including intervention studies demonstrating that dietary improvement produces measurable improvements in mental health outcomes independent of other treatments. The implication for brain health is significant: what you eat not only fuels your brain.

The sugar story, personally

I want to be specific about what this looked like in my own life, because the abstract language of glycaemic indices and inflammatory markers can make dietary neuroscience feel distant from the daily reality of a pressured professional reaching for the nearest available comfort.

There was a period in my academic career when I noticed a pattern that, once I saw it, I could not unsee. After difficult meetings, budget negotiations, manuscript rejections, and tense conversations with collaborators, I would return to my office and reach for something sweet, like chocolate or a snack bar with enough wholesome words on the packaging to feel nutritious. It was not hunger, and the reach was automatic, and the relief that followed it was immediate and real.

What I was doing was using sugar as a stress regulator. Stress elevated by the difficult meeting narrowed my attention to the threat and created a state of low-grade discomfort. Sugar, specifically its activation of dopamine release, provided rapid, reliable relief. The loop was efficient: discomfort, reach, relief. Repeat it often enough, and it becomes circuitry. The brain learns to anticipate the reward and initiates the seeking behaviour before conscious deliberation can stop it. It is reinforcement learning, using the same mechanism by which the brain acquires habits, operating in this case on a substrate of stress hormones and dopamine. The research on sugar and reward pathways shows that the fructose component of sucrose activates the brain in ways that share mechanisms with addictive substances, that repeated exposure reduces the sensitivity of dopamine receptors, and that the blood glucose crash that follows a spike increases stress reactivity and, in turn, makes the next stressful event more dysregulating, which makes reaching for sugar more likely.

When I reduced my sugar intake, I did not aim for perfection; I removed one item at a time. One week without after-dinner chocolate, then one week without trail mix at my desk. The first few days of each change were uncomfortable in a specific way: my brain had learned to expect a reward at a certain time in a certain context, and its absence registered as a mild but persistent unease. That discomfort was not a sign that something was wrong. It was the brain circuitry beginning to adjust to conditions that no longer reliably delivered the anticipated reward. It was, in the most literal sense, rewiring.

Within two weeks of each change, the craving somewhat diminished. Within a month, the automatic reach stopped, and within three months, energy was substantially more stable across the day, and the afternoon cognitive flatness I had attributed to 'the hard work' had largely resolved. Neuroplasticity is cumulative, quiet, and proportional to the consistency of the new input.

What the brain actually needs from food

The dietary pattern most consistently associated with brain health across the large bodies of research is one that will surprise almost no one: predominantly whole foods, high in fibre, diverse in plant matter, rich in omega-3 fatty acids, low in ultra-processed ingredients, and moderate in refined carbohydrates.

The Mediterranean dietary pattern, characterised by vegetables, legumes, whole grains, fish, olive oil, and moderate amounts of fermented dairy, has the strongest evidence base for cognitive protection and mental health benefits. The MIND diet, a variant developed specifically to target cognitive resilience, has been associated with significantly reduced rates of cognitive decline in longitudinal studies. Neither requires perfection, expensive ingredients, or the kind of dietary rigidity that, as we now understand, is counterproductive.

The three most impactful dietary changes for cognitive performance, based on research rather than industry, are: reducing intake of ultra-processed foods, stabilising blood glucose by including protein, fat, and fibre at each meal, and increasing dietary diversity, particularly plant diversity, to support microbiome health. These changes do not require a dietary philosophy. They require incremental improvements in decision-making, consistently over time.

In practice: Before any meal, ask one question: Does this meal contain protein, fat, and fibre? If yes, it will produce a slower, more stable glucose response than a carbohydrate-dominant meal, and your cognitive performance over the next two to three hours will reflect that stability.

Sleep: The Brain's Non-Negotiable Maintenance Window

If I were asked to identify the single most common modifiable contributor to poor performance, emotional dysregulation, and reduced resilience among professionals, it would not be diet, exercise, or stress management. It would be sleep, specifically, the chronic, normalised, socially sanctioned under-sleeping that characterises high-achievement professional culture and that most people have learned to regard as an unavoidable cost of ambition. Sleep is not rest and is an active, structured biological process with distinct phases, each performing functions that waking life cannot replicate. Understanding what actually happens during sleep, not the generic instruction to get more of it, but the specific biological work that occurs during each phase, is what makes the case for sleep protection compelling in a way that no amount of general advice achieves.

What happens when you sleep

Sleep is not a single, uniform state but a structured cycle made up of four distinct stages that repeat throughout the night. The first three stages are non-REM sleep. Stage 1 (N1) is the lightest phase and occurs as you drift into sleep. Stage 2 (N2) involves deeper relaxation, with a drop in body temperature, slower breathing and heart rate, and reduced muscle activity. Stage 3 (N3), often called deep sleep, is the most restorative phase, supporting physical recovery and growth. It is also during this stage that the brain's glymphatic system becomes highly active, clearing metabolic waste and toxins that accumulate during waking hours. The fourth stage is REM sleep, when brain activity increases, most dreaming occurs, and the body becomes temporarily immobilised. A typical night involves four to six of these cycles, each lasting about 90 minutes, although the length and composition of cycles change across the night and vary between individuals.

When thinking about healthy sleep, it is common to focus only on duration, but quality is equally important. Each stage plays a distinct role in restoring the body and brain, and disruptions to the cycle can affect how refreshed a person feels on waking. Research by Matthew Walker has highlighted that both deep and REM sleep are essential for memory consolidation, emotional regulation, and long-term brain health. Understanding sleep architecture helps explain why conditions such as insomnia or obstructive sleep apnoea can have such a significant impact, as they interfere with the natural progression through these stages. Good sleep is therefore not just about getting enough hours, but about moving through the full cycle in a way that allows the body and mind to recover effectively.

Sleeping six hours instead of eight is not getting 75% of the benefit. It is getting only a fraction of the benefit, because the functions that occur in the final two hours of a full sleep cycle are among the most critical and are the first to be lost.

17 hours of wakefulness produces cognitive impairment equivalent to a blood alcohol level of 0.05%, the legal driving limit in most Australian states

Williamson et al., 2000 Road Safety Research Report, Department of Transport and Regional Services Australian Transport Safety Bureau

The cost of chronic short sleep

After approximately seventeen hours of continuous wakefulness, cognitive performance degrades to a level equivalent to a blood alcohol concentration of 0.05 per cent, the legal driving limit in Australia. Most professionals are operating in a state of accumulated sleep debt that research has shown impairs performance in ways that the impaired person consistently underestimates: a characteristic feature of sleep deprivation is reduced insight into one's own degraded capacity.

The specific cognitive functions most vulnerable to sleep deprivation are, predictably, the ones most valuable in professional settings: sustained attention, working memory, emotional regulation, creative problem-solving, and the capacity toward social judgement. These are precisely the capacities that AI cannot replicate and that the AI era demands most from its human participants. Protecting sleep is protecting professional capability.

Emotional regulation, the capacity to respond to difficult situations proportionately rather than reactively, is particularly sleep dependent. The prefrontal cortex's ability to modulate amygdala reactivity is one of the first casualties of insufficient sleep, which is why sleep-deprived professionals tend to experience more intense emotional reactions to relatively minor frustrations, make poorer decisions under social pressure, and find it harder to be empathetic than effective leadership and professional relationships require.

Sleep hygiene is not about wind-down rituals.

The term sleep hygiene has become associated with lavender candles and blue-light glasses that it has lost much of its scientific content. What the evidence actually shows is more structural and more actionable than any ritual.

Sleep pressure, the biological drive to sleep, is regulated by two systems. The circadian system, driven by light exposure and body temperature, establishes the timing of sleep. The homeostatic system, driven by adenosine accumulation during wakefulness, regulates the intensity of sleep drive. Both systems can be protected or reduced by specific behaviours.

Regular Sleep Schedule

The National Sleep Foundation reports that maintaining a regular sleep schedule by going to bed and waking up at consistent times each day is a simple yet powerful way to support both mental and physical health. When sleep becomes a priority, it is easier to manage stress, maintain focus, and support overall well-being, including heart health and daily performance. Consistency helps regulate the body's internal clock, making it easier to fall asleep and wake feeling refreshed. Practical steps such as setting bedtime reminders, using a morning alarm, and following a calming wind-down routine each night can reinforce this rhythm. Over time, these habits improve sleep quality, reduce the impact of sleep debt, and help the body and brain function at their best.

Light exposure is the primary circadian signal. Morning bright light, ideally natural light, within 30 minutes of waking advances the circadian phase and strengthens the sleep-wake rhythm. Evening screen light delays it, and the blue-light filtering advice is correct as far as it goes, but incomplete: the more important intervention is the one earlier in the day, when morning light sets the biological clock that will determine sleep onset twelve to sixteen hours later.

In practice: Set one wake time and hold it for 7 days a week for 4 weeks. Before adjusting anything else about your sleep, before the supplements, the apps, the rituals, try this. It is the foundational intervention on which all other sleep improvements depend.

Energy: The Overlooked Dimension of Cognitive Capacity

Energy is the dimension of human performance most frequently discussed and least carefully examined. We speak about energy levels the way we speak about the weather, as something that happens to us, that varies in ways we observe but do not control, that is restored by coffee and depleted by demanding days.

Energy, in the neurological sense, is cognitive bandwidth. It is the resource that determines whether attention can be sustained or fragments, whether emotional regulation is available or reactive responses take over, whether long-term thinking is accessible or short-term urgency dominates. It is the actual physiological capacity of the nervous system to sustain the quality of function that professional life demands.

This capacity is determined primarily by four variables: sleep quality, movement frequency, nutritional pattern, and recovery, as well as the deliberate restoration of nervous system resources between periods of activation. We have now discussed three of these. The fourth, recovery, is the variable most routinely sacrificed in high-performance work settings, and its sacrifice has consequences that compound over time.

The inflammation tax

Chronic low-grade inflammation is the common pathway through which poor sleep, sedentary behaviour, ultra-processed diet, and chronic stress all converge to impair cognitive function. It is worth understanding how this works because inflammation is often discussed as if it were primarily a physical health concern and relevant to cardiovascular disease and metabolic syndrome, but separate from the cognitive and emotional domains that professionals care most about.

Inflammatory cytokines, the signalling molecules released during systemic inflammation, cross the blood-brain barrier and act directly on brain tissue. The subjective experience of elevated neuroinflammation includes reduced motivation, impaired concentration, emotional flatness, increased sensitivity to social threat, and brain fog, the sense that thinking requires unusual effort, and the clarity that formerly felt has gone.

The inflammation tax is cumulative and largely invisible. A professional operating under chronic neuroinflammation is performing below their capacity every day, in ways they may have normalised over months or years as simply 'how I am now.' The research suggests that this normalisation is both common and unnecessary: dietary improvements, movement, and sleep protection produce measurable reductions in inflammatory markers within weeks, with corresponding improvements in brain performance that many people describe as a return to the person they used to be.

Recovery is not the opposite of performance.

Perhaps the most important reframe in this chapter is that recovery is not the absence of performance. It is the mechanism through which performance becomes sustainable.

The professional culture that treats rest as a reward for completed work, that regards a full calendar as evidence of value, and that equates constant availability with professional commitment is not a high-performance culture. It is a culture systematically depleting the brain capital resources on which high performance depends.

Research on elite performance across domains such as sport, surgery, music, and chess consistently identifies recovery as a non-negotiable component of sustained excellence. Deliberate practice, the intensive, focused effort that builds expertise, is only effective when alternated with periods of rest that allow memory consolidation, neuroplasticity, and motivation, thereby making the next period of effort productive. Performance without recovery does not remain flat. It degrades.

The most productive professionals are not those who work the longest hours, but those who protect the biological conditions that enable their work. They understand that these conditions are built during both rest and work periods.

In an AI-augmented professional environment, this matters more than ever. The cognitive tasks that AI handles well are not particularly dependent on the quality of the human operator's biological state. But the tasks that remain distinctively human are exquisitely sensitive to that state. A tired, inflamed, sleep-deprived professional interacting with their best AI tools produces output constrained by the human-in-the-loop's brain and mental health.

In practice: Identify one recovery practice you are currently treating as optional, a lunch break, a non-negotiable end time, a screen-free evening, and make it happen for thirty days. Track your cognitive quality in the mornings that follow. The change will not be dramatic. It will be quiet, cumulative, and after thirty days, unmistakable.

The Body as Foundation, Not Afterthought

The premise of this chapter is simple, and the evidence for it is now overwhelming: the quality of the brain available for the work of your life is determined primarily by the quality of the biological conditions in which it operates. Not entirely; genetics, history, and circumstance all play roles that biology alone cannot override. But within your influence, the four fundamentals of movement, food, sleep, and energy restoration are the highest-return investments available.

This is the human upgrade. It is not a new productivity system, a cognitive enhancement supplement, or a technology-mediated optimisation of working patterns. The upgrade that the research consistently and repeatedly identifies is the restoration of the biological conditions that the human nervous system evolved to require, and that modern professional life has normalised the erosion of.

Artificial intelligence will continue to expand what is cognitively possible in information processing. It will not metabolise your cortisol or produce your BDNF. It will not consolidate your memories, recalibrate your amygdala, or restore the prefrontal capacity that sustained pressure depletes. Those processes are biological and require biological inputs that no technology can substitute for.

In a world where AI handles more and more of the cognitive effort, professional advantage shifts to whoever brings the clearest, most regulated, most rested human brain to the work that remains. That brain is built in the gym, the kitchen, the bedroom, and the deliberate pause between demands.

The more complex and demanding the external environment becomes, the more essential the internal foundations are. This is not a retreat from ambition but a prerequisite for sustaining it.

My mother walked for two hours before dawn every day of her adult life without knowing the word exerkine or understanding BDNF. She built her brain, one dark morning at a time, through a practice that science has since validated with considerable precision. What she could not give herself and what we can give ourselves, with the knowledge now available, is the full picture: the sleep she should have protected more carefully, the glucose stability that would have quieted the worry that ran beneath her discipline, the recovery that would have made her extraordinary endurance sustainable rather than quietly costly.

We have the evidence she did not. The question is whether we use it. Start with movement, sleep, food, and recovery, not all at once, since change built on overwhelm does not last. Make one change for 30 days until it becomes a condition, then add the next. The brain you are building is the one you will think with for the rest of your life.

CHAPTER SIX

The First Hour

How you start the day and small daily shifts rewire your brain

Let me tell you about a moment when I was not expecting to change my mind about anything.

This is the argument for the first hour: not that morning is inherently sacred, not that any particular ritual is required, but that the brain state in which you begin your day has downstream consequences that most people significantly underestimate. And that the most reliable way to begin from a regulated, attentive, purposeful state rather than a reactive one is to protect the first hour from the thing that most reliably prevents it: the phone.

The moment you look at your phone, you have handed the direction of your attention to everyone who sent you something overnight. Everything that follows, the emotional activation of the news, the low-grade urgency of unread messages, the comparison triggered by social media, enters before you have had the opportunity to establish your own baseline.

The practice of protecting the first minute to an hour from screens is not technophobia. It is becoming aware of where our eyeballs are landing and what they are being shown. Checking the phone immediately after waking activates the brain's stress centres before it has had the opportunity to start the day in a calm, reflective way, which later in the day diminishes under pressure.

The morning routine that works

The morning routine I have built over the years started with a single practice, taking in a panoramic view and thinking of three things I was grateful for. The conversation that mattered yesterday, the fact I can wiggle my toes, the way the light looked this morning. The colleague who said the thing that shifted something. The brain mechanisms of gratitude practice and taking a wide view help activate the medial prefrontal cortex, promote the release of dopamine and serotonin, and shift the default mode network's attentional bias toward more positive than negative news and aspects of our lives.

Water before anything else, as the brain has been without fluid for seven or eight hours and cognitive performance, including attention and working memory, is measurably impaired at levels of dehydration well below the threshold of felt thirst. Movement, even briefly. Five to ten minutes of gentle stretching or walking, nothing that requires equipment, a gym, or the activation of the discipline that early mornings frequently fail to produce. The purpose is not only fitness but also to increase blood flow and cerebral circulation, to generate the initial exercise signal that tells the brain it is awake and operational, and to produce the first small dopamine release that makes every subsequent morning practice slightly easier.

I was preparing to give a keynote address at a conference, a setting I have been in hundreds of times, with material I knew well, in front of an audience that was not hostile. By any reasonable measure, it should not have been a source of significant anxiety. And yet, standing in the corridor outside the auditorium thirty minutes before I was due to speak, I was aware of a familiar unwelcome physiology: the signature of the stress response, running quietly in the background, doing nothing useful.

I had been researching the brain for years and understood the connection between the brainstem's breathing centre and the amygdala is bidirectional and that the amygdala signals the breathing centre when it detects threat, producing the shallow, rapid pattern we call anxious breathing, but that the pathway works in both directions, meaning that deliberately changing the breath can send signals back to the amygdala. That afternoon, I sat down in the corridor, set a timer for five minutes, and breathed deliberately and slowly, with the full attention I had been recommending to others for years but applying to myself only sporadically.

Within two minutes, something altered dramatically, the anxiety dissolves, and you walk on stage feeling ready.

Of all the tools for brain health that three decades of neuroscience research have made available, breath may be the most underused, not because it is hard to access but because it is so ordinary that we keep overlooking it.

The Bidirectional Circuit: Mind to Body

The brainstem contains a cluster of neurons called the preBötzinger complex, which generates the automatic rhythm of breathing. Research led by Dr Jack Feldman at UCLA established that this breathing centre is not simply a mechanical regulator; it has direct neural connections to the locus coeruleus, the brain's primary noradrenaline centre, and, through it, to the amygdala and the broader arousal system. This means that breathing patterns and emotional states are not merely correlated. They are causally connected, in both directions.

The bidirectionality is what makes breath a usable tool rather than merely a symptom. Because the connection runs both ways, deliberately slowing and deepening the breath sends signals back along the same circuit. Slow, rhythmic, diaphragmatic breathing communicates with the brain's stress circuits through direct neural input, allowing the assessment of threat to be revised. That the body is not, in fact, preparing to flee, and the system can stand down.

And it means that breath is the only component of the autonomic nervous system that is simultaneously under automatic control and available to voluntary direction. But you can control your breath, and through your breath, you can influence all of those things.

In practice: Before any situation you anticipate finding stressful, take 2 minutes and breathe out longer than you breathe in. In for 4, out for 6 or 8. The specific counts matter less than the ratio. Most people notice a shift within three or four cycles. You are not eliminating the stress response. You are preventing it from running unchecked.

For sustained regulation: box breathing

Box breathing, four counts in, hold for four, four counts out, hold empty for four is the technique that has been adopted by military and emergency services personnel for use under high-stakes conditions, as it works reliably, can be practised anywhere without drawing attention, and produces a level of nervous system regulation that is useful precisely in the conditions where most techniques fail because the person cannot remember to use them or cannot slow down enough to apply them. The four-sided symmetry, or box, provides a structure for the mind to focus on, which prevents ruminative thinking that tends to amplify stress responses by keeping the prefrontal cortex occupied with something simple and concrete. Five minutes of box breathing produces a measurable shift in stress and heart rate variability.

In practice: Four counts in through the nose. Hold for four. Four counts out through the mouth. Hold empty for four. Repeat for five minutes. Set a timer so you don't check the clock. The first minute is often uncomfortable, as the mind wants to do something else. By the third minute, the shift is usually perceptible. By the fifth, most people are in a qualitatively different physiological state than when they started.

Habit formation and Neuroplasticity

The principle of habit formation is the practice of attaching a new practice to an existing one. The practice borrows the contextual cues and neural triggers of the established one, which dramatically reduces the activation energy required to initiate it.

If you normally reach for your phone as soon as you wake, try looking out the window and taking in a wide panoramic view instead. If you already make coffee every morning, try a breathing practice while it brews. If you've already had a glass of water, add two minutes of stretching immediately afterwards. If you already have a moment before getting out of bed, use it for three gratitude's before your feet touch the floor. The sequence does not need to be elaborate, just consistent enough that the practices become what the neuroscience of habit formation calls automatised, that is, no longer requiring deliberate decision and becomes simply what happens in the morning.

Start with the smallest one that requires the least effort and produces the most clearly perceptible effect. For most people, taking in a panoramic view and practising gratitude is that one, as it takes one or two minutes, requires nothing, and produces a detectable shift in mood within the first week of consistent practice by not looking at your phone as the first thing you do. Add the next element only when the first has become something you do rather than something you are trying to do.

When the routine breaks

The morning routine is often disrupted by children, under- or oversleeping, and travel. You will have early commitments that compress the available time to nothing. You will have nights from which you emerge too depleted to do anything with intention. These are not mistakes, but the ordinary texture of a life that is not organised around a morning routine.

The minimum viable version for difficult mornings is the panoramic view, three gratitudes and slow breaths. Not the full effect of the complete practice, but enough to prevent the complete abandonment of the practice, which is what matters, because it is the pattern that produces the compound effect, not any individual morning.

Missing one day is not failure, and the response to missing most days is not renewed resolution but honest examination of what the barrier actually is, which is usually that the routine is too elaborate to be sustainable under the ordinary conditions of the life you are actually living, rather than the life you imagined you would have when you designed it.

The morning practice that you will actually do consistently is worth more than the ideal practice you will abandon in the second week.

The Compound Return

The reason morning practices matter is not that any individual morning produces a measurable change in the brain. It is the accumulation of mornings that does.

Neuroplasticity is a slow process, and neural pathways are strengthened by awareness and attention to where we focus, by breathing through repetition over weeks and months. This is neuroplasticity working exactly as it is designed to work. Not through the perfect morning but through the ordinary one, returned to again, offered what it needs, and trusted to do the slow work of building a brain that can bear what the day brings and recover from what it cannot. You can start tomorrow, not with everything at once but with one breath, taken with intention, before you look at your phone. That is where the compound return begins.

CHAPTER SEVEN

The Power Between Us

Why is connection the ultimate performance advantage

There is a moment in every relationship, every mentoring conversation, every genuinely important meeting, when something shifts that no transcript could capture.

I first heard about the research on interbrain synchrony from Dr Joy Hirsch while interviewing her for the Thriving Minds podcast. Two people were sitting across from each other, talking about something that mattered to both of them. Her team were recording their brain activity simultaneously, and on the screen, their neural patterns were beginning to align in a rhythmic, reciprocal coordination that looked like two instruments finding the same key.

This is called inter-brain synchrony. It occurs when two or more people are genuinely engaged with each other, emotionally attuned, mutually focused. Their neural oscillations begin to couple, particularly in the prefrontal and temporal regions associated with empathy, language processing, and social cognition.

In an era when artificial intelligence can analyse, synthesise, and generate at extraordinary speed, this capacity, the biological capacity of one human nervous system to regulate and be regulated by another, will be the foundation of every meaningful professional relationship, every act of genuine leadership, and every collaboration that produces something none of the participants could have achieved alone.

This chapter is about what happens between us. Why it matters, what is threatening it, and how to protect it as the most valuable human capability we possess.

The Biology of Connection

Human beings are not built to function in isolation, and the brain develops in relationship. From the first moments of life, the neural architecture that governs emotional regulation, stress tolerance, and social cognition is shaped by the quality of interpersonal interaction. These are called serve-and-return exchanges between infant and caregiver; the attuned gaze, the voice, and the physical holding are not merely comforting. They build the neural circuitry that allows a human being to manage distress, read social situations, and form a secure, safe attachment, from which exploration, learning, and risk-taking become possible.

This architecture does not become irrelevant in adulthood. It becomes the operating system on which professional performance runs. When a colleague walks into a room, and you can sense something is wrong before they speak, you are reading micro-signals, facial expressions, shifts in body posture, and vocal tone that your brain processes largely below conscious awareness. When a skilled clinician sits with a distressed patient, and the patient begins to settle, that is not simply good bedside manner. The clinician's regulated nervous system is physiologically influencing the patients.

> *Connection is not the soft side of performance. It is the biological infrastructure on which performance depends.*

Why Screens Cannot Replace Presence

We live in an era of extraordinary connectivity and escalating loneliness. Digital communication transmits information but does not transmit the full bandwidth of human presence. This matters more than most professionals realise. When you are in a room with someone, your brain is processing an extraordinary volume of social information, most of it below the threshold of conscious awareness. It is tracking eye movement, intonation, gestural timing, and the synchronisation of your physiological states. It is the evolved mechanism through which human beings establish trust, detect deception, and create the felt sense of being understood.

Remote communication strips much of this away. The cumulative effect of predominantly screen-mediated interaction over weeks, months, and years is a subtle yet measurable erosion of the interpersonal regulatory capacity our nervous systems depend on.

The research on loneliness bears this out. Loneliness is not merely unpleasant; it is a physiological stressor, and chronic loneliness activates the same inflammatory pathways as chronic physical stress and is associated with impaired immune function, accelerated cognitive decline, and increased cardiovascular risk. The epidemiological data now suggest that severe social isolation carries health risks comparable to smoking fifteen cigarettes a day.

> *We are the most digitally connected generation in history, and among the loneliest. The technology that promised to bring us together has, in many cases, replaced the depth of connection with the breadth of contact.*

For professionals navigating the AI transition, the capacity for deep, regulated, face-to-face connection is precisely what AI cannot replicate. It is also the capacity most at risk in environments increasingly mediated by screens, remote work, and algorithmic interaction.

Co-Regulation: The Hidden Professional Skill

There is a term in developmental psychology that I believe belongs in every professional capability framework: co-regulation.

Co-regulation describes the process by which one person's nervous system helps stabilise another's. It is the mechanism through which a calm parent settles an agitated child, a composed leader steadies an anxious team, and a skilled therapist creates the conditions in which a client can access difficult material without becoming overwhelmed.

It is not a technique, and it requires, as a prerequisite, that the person doing the regulating has a nervous system that is itself reasonably well regulated. You cannot lend calm you do not possess. You cannot offer a steadying presence amid internal chaos.

This is why the earlier chapters of this book, on sleep, movement, nutrition, and morning routine, are not separate from this chapter. They are foundational to it. The capacity to co-regulate depends on the biological infrastructure that those practices build and maintain. A sleep-deprived professional with an overstimulated nervous system and chronic low-grade stress activation cannot provide the regulatory presence that their clients, patients, students, or colleagues need, regardless of their technical expertise.

Consider what this means in practice:

For lawyers: when a client is making decisions in a state of high emotional activation, during a custody dispute, a criminal proceeding, a corporate crisis, the lawyer's capacity to remain regulated, to hold the space without being pulled into the client's distress, directly influences the quality of the decisions that follow. This is not emotional detachment. It is regulated engagement, and it is a skill that AI cannot provide.

For educators: a classroom is a regulatory environment. The teacher's nervous system sets the baseline for the room. A dysregulated teacher produces a dysregulated classroom, not through any failure of content delivery, but through the biological mechanisms of social contagion. Emotional states are literally infectious, transmitted through facial expression, vocal tone, and postural cues that students process below conscious awareness.

For leaders: the most consequential thing a leader does in a crisis is not the strategy they choose. It is the state they bring to the room. A regulated leader creates the neurobiological conditions under which the team can think clearly, collaborate effectively, and tolerate the uncertainty that crisis demands. A dysregulated leader, regardless of their strategic brilliance, activates threat responses in their team that narrow thinking, reduce creativity, and undermine the very collaboration the moment requires.

> *The most undervalued professional skill in the AI era is not a skill at all. It is a state: the capacity to be regulated enough to regulate others.*

The Science of Trust

Trust is not an abstraction. It has specific neural correlates and specific conditions under which it develops or erodes.

Oxytocin, often simplified in popular accounts as the bonding hormone, plays a role in trust formation, but the neuroscience is more nuanced than the headlines suggest. Oxytocin facilitates in-group bonding and social approach behaviour and modulates the brain's social-evaluation circuitry, making us more attuned to social signals and more responsive to cooperative cues from people we regard as safe.

Trust develops through repeated experiences of reliability. Not through declarations of intent, not through mission statements, but through the accumulation of micro-interactions in which what was promised is delivered, in which vulnerability is met with respect rather than exploitation, and in which the social signals of safety are consistent over time.

This has direct implications for professional environments. Organisations that espouse trust while creating conditions of surveillance, unpredictability, or punitive accountability are working against the neurobiology of the people within them. The brain is exquisitely sensitive to discrepancies between stated intent and experienced reality. When those discrepancies are chronic, the stress-response system activates, interpersonal vigilance increases, and the collaborative capacity that trust enables is quietly undermined.

In an AI-augmented workplace, where roles are shifting and certainty is scarce, the conditions for trust are both more fragile and more important than ever. Professionals who can build and sustain trust, through consistency, through presence, through the regulated engagement that signals genuine safety, hold a capability that no algorithm can replicate or replace.

Quality Over Quantity: What the Research Shows

Robin Dunbar's research on social network size suggests that the human brain can maintain approximately one hundred and fifty meaningful social relationships, with far fewer, roughly five, in the innermost circle of deep trust and emotional intimacy. These numbers are not arbitrary. They reflect the cognitive and emotional bandwidth of the human brain, and specifically the capacity of the prefrontal cortex to track the complex social information that genuine relationships require.

Social media has exploded the number of connections available to us without expanding the neural architecture that manages them. The result, for many professionals, is a wide but shallow social network, rich in contacts but poor in the kind of deep, regulated, reciprocal relationships that the nervous system actually needs.

The loneliness research confirms this. Social isolation and loneliness are not the same thing. A person with a large social network can be profoundly lonely if none of those connections reaches the depth at which genuine co-regulation occurs. Conversely, a person with very few relationships can be deeply nourished if those relationships are characterised by presence, trust, and connection.

What I have found in my own life, and what I hear consistently from the professionals I work with, is that the reduction in social breadth, the deliberate choice to invest in fewer but deeper relationships, produces a quality of connection that is qualitatively different from anything that social volume provides. It is the difference between being known by many and being understood by someone. And it is the understanding, not the knowing, that the nervous system can use.

> *The brain does not need more connections. It needs deeper ones, and in the attention economy, depth of relationship is the scarcest resource and the most protective.*

Connection in the Age of AI

Artificial intelligence will continue to improve at tasks that involve information processing, pattern recognition, and language generation. What it will not do, because it cannot do, is regulate a nervous system in the same way our biological quantum human being can.

When you sit with someone who is frightened, and your calm presence helps them think more clearly, that is a biological and social process. When you lead a team through uncertainty and your regulated state creates the conditions for them to collaborate rather than retreat into self-protection, that is neuroscience in action. When you mentor a junior colleague, and they begin to develop the professional judgement that comes from observing how a seasoned practitioner thinks, weighs evidence, and makes decisions under pressure, that learning happens through the full bandwidth of human presence in ways that no recording, no transcript, and no simulation can replicate.

The AI transition will reshape professional environments in ways we cannot yet fully anticipate. Some of those changes will be extraordinarily productive. Others will require vigilance. But the thread running through this book is that the human capacities that matter most, the ones that determine whether we thrive or merely survive, are biological, relational, and built through the daily practice of showing up with presence, intention, and a nervous system capable of meeting whatever the moment requires.

Every genuine conversation, every moment of undivided attention offered to another person, every decision to put the screen down and be present, these are not small acts. They are the conditions under which the most powerful human capability we possess, the capacity to regulate and be regulated by each other, is maintained and strengthened.

Building Connection: What the Research Recommends

Prioritise face-to-face interaction. Video calls are better than text, and text is better than silence. But nothing provides the full regulatory bandwidth of being physically present with another person. Where possible, protect time for in-person connection, particularly with the people whose relationships matter most to your wellbeing and professional practice.

> **In Practice:** *Identify one professional relationship that you have been maintaining primarily through screens. In the next fortnight, arrange to meet that person in person. Notice what changes in the quality of the conversation when you are physically present.*

Invest in depth over breadth. The temptation, particularly for professionals whose networks are large, is to maintain contact with many people at a shallow level. The research consistently shows that fewer, deeper relationships are more protective for mental health and more productive for professional collaboration than extensive but superficial networks.

> **In Practice:** *Choose three relationships that you want to deepen over the next three months. Commit to one genuine, unhurried conversation with each of them this week, a conversation that is not about logistics or tasks, but about how they are actually doing.*

Regulate yourself first. Co-regulation begins with self-regulation. The practices outlined in earlier chapters, movement, food, sleep, the first hour, deliberate recovery, and now social connection, help to build a well-regulated nervous system, which is the most valuable thing you bring to every relationship.

> **In Practice:** *Before your next important meeting or conversation, take ninety seconds to regulate your breathing. Three slow breaths, each with an exhale longer than the inhale. This is preparation and readying your nervous system to offer the regulated presence that every meaningful interaction requires.*

Listen with your body, not just your mind. Genuine listening involves more than processing words. It involves orienting your attention, posture, and presence toward the other person in a way their nervous system can register. This means maintaining eye contact, remaining still, and deliberately suspending the urge to formulate your response while they are still speaking.

> **In Practice:** *In your next conversation, practise listening without planning what to say next. Simply attend, let the other person finish, and allow a moment of silence before you respond. Notice what happens to the quality of the exchange.*

The power between us is not given. It is practised. And in an age when machines are becoming more capable with every passing month, the human capacity to be genuinely present with another human being is not diminishing in importance. It is becoming the most consequential thing we do.

C H A P T E R E I G H T

The Stories We Believe

Bias, blind spots, and the courage to rethink everything

At the beginning of my career, I held a belief that I did not examine very carefully. It lived below the level of deliberate thought, as most professional assumptions often do. I believed that mental health disorders were primarily problems of chemistry and that the most effective solutions would come through pharmacology. It was not a position I consciously defended; it was simply embedded in my identity as a pharmacist and emerging neuroscientist. Questioning it would have felt, in some quiet way, like questioning myself.

I understood, intellectually, that medications for depression, anxiety, and related conditions were legitimate clinical tools supported by evidence. I could describe their pharmacological mechanisms in detail, knew how selective serotonin reuptake inhibitors functioned, how dopamine circuits modulated, and how receptor binding translated into symptom change. That knowledge felt solid and dependable, giving structure to suffering.

What I had not fully examined was the deeper question of cause. I had not yet asked whether altered chemistry was the origin of the problem, or whether it might be a downstream expression of something more complex.

The key moment occurred when I moved to the University of California, San Francisco. At UCSF, I was surrounded by some of the world's leading neuroscientists. It was there that I encountered, in a far more rigorous way, the science of neuroplasticity. The brain was not fixed; it was not a static organ determined solely by neurotransmitter deficits. It was dynamic, adaptive, and continuously reshaped by experience.

Neuroplasticity was not a motivational concept; it was measurable biology where synapses strengthened or weakened based on use. Circuits reorganised in response to stress, learning, and environment. Early life experiences left structural and functional signatures in the brain, and repeated patterns of thought and behaviour altered connectivity over time.

This understanding shifted something fundamental in me. If the brain could change in response to adversity, then perhaps the symptoms my sister was experiencing were not simply chemical imbalances to be corrected. Perhaps they were adaptations to stress, trauma, or a lack of safety and love. Perhaps medication addressed part of the picture, but not the whole.

It took that encounter with plasticity science to move my thinking from symptom management toward systems understanding. I began to see mental health not only as a matter of neurochemistry, but as a matter of networks, experiences, and environments interacting across time. That shift did not make medication irrelevant. It made it incomplete, and as a scientist, I had to understand this deeply.

I want to offer an invitation rather than a judgement: my resistance was not scientific. I had been telling a story about what genuine healing looked like, and that story had a shape that medication did not fit neatly into. The story was not without basis, as lifestyle intervention is powerful, and the biological fundamentals of the previous chapter are as well. But the story had become a filter, and filters, by their nature, remove things that do not belong to the pattern they are designed to recognise.

The shift I made was not to a new certainty but rather to a new curiosity. I began asking, when someone described a mental health struggle, not: have they tried the approaches I value most? But: what does this specific person's brain and mind actually need, in this specific moment, to create the conditions in which their own capacity for growth and healing can operate? Sometimes the answer is therapy; sometimes movement; sometimes connection and community; sometimes healthy food and sleep. And sometimes, it is medication, used thoughtfully, as a bridge to the conditions in which everything else becomes accessible.

The most important shift I made was not changing what I believed about mental health, treatment and healing. I became aware that I had been holding a belief and was failing to ask whether it was serving the people it was supposed to serve.

I share this not to advocate for any particular clinical approach, which is a conversation between each person and the professionals who know their specific situation, but because this experience revealed to me something about the nature of belief that I have not stopped thinking about since. The story I was telling about medication felt like professional judgement. It was also, to a degree, that I had not recognised, becoming another professional identity. And professional identity, it turns out, is one of the most powerful and least examined filters through which we process new information.

This chapter is about that filter. About the neuroscience of how belief shapes what we perceive, what we accept, and what we resist. About the specific cognitive patterns, unconscious incompetence, confirmation bias, and identity-protective reasoning that make even intelligent, well-intentioned people systematically blind to the limits of their own understanding. And about what it actually takes, neurologically and psychologically, to update a belief that has become part of who you think you are.

The Believing Brain

The popular model of human cognition imagines a mind that receives information, evaluates it objectively, and updates its understanding accordingly. This model is appealing and almost entirely wrong.

The brain is not a neutral information processor. It is a prediction machine and a system that has evolved to build models of the world and to use those models to anticipate what will happen next, rather than to register what is happening now. The models it builds are efficient precisely because they are conservative: they update slowly, they resist contradiction, and they prioritise coherence over accuracy. A brain that revised its understanding of the world every time it encountered new information would be unable to function, and the cognitive cost of perpetual revision would be prohibitive. The brain's conservatism is, in most circumstances, adaptive.

The neurological mechanism of belief maintenance involves two distinct systems working in tandem. When information matches existing beliefs, reward circuits activate. The nucleus accumbens, the identical structure involved in the dopamine responses discussed in earlier chapters, registers coherence as a form of reward. Being right, or more precisely, encountering information that confirms what you already think, produces a mild but genuine neurochemical pleasure. This is why consuming media that confirms our existing views feels satisfying in a way that is disproportionate to any practical benefit.

When information contradicts an existing belief, the response is qualitatively different. The stress response initiates, and the experience of cognitive dissonance, the uncomfortable tension of holding two contradictory positions simultaneously, is not simply psychological, but can become a stress signature. And the brain, which has learned through long evolutionary experience that stress is a signal to resolve threat rather than to tolerate it, moves immediately and largely unconsciously to restore coherence.

The brain resolves the discomfort of cognitive dissonance not by examining which belief is more accurate, but by whichever route requires the least cognitive effort. Most of the time, that route dismisses the new information rather than updating the old belief.

The consequence is a cognitive system that, under normal operating conditions, is structurally biased toward preserving existing beliefs. Not because we are irrational, but because the mechanisms that govern belief maintenance evolved in an environment where the speed of response mattered more than the accuracy of updating. A belief system that held firm under most challenges was more useful, evolutionarily, than one that revised itself carefully but slowly in response to evidence.

We are now living in an environment where this tendency has become a liability. The pace of change in professional knowledge, the breadth of expertise required to handle complex challenges, and the acceleration of AI-generated information create conditions in which the accuracy of our understanding matters more than the speed with which we defend it. The brain's preference for logicality and accuracy, once adaptive, is now one of the main obstacles to the kind of persistent learning that professional life increasingly requires.

What We Do Not Know We Do Not Know

In 1999, the psychologists David Dunning and Justin Kruger published a paper titled "Unskilled and Unaware of It," which has since become one of the most widely cited findings in social psychology. The study showed that people with limited expertise in a domain consistently overestimate their competence in that domain. It describes a pattern that is both uncomfortable and, once understood, genuinely liberating.

The skills required to perform well in a domain are, to a significant degree, the same skills required to evaluate performance accurately. When you lack expertise in an area, you lack not only the capacity to perform well, but also the metacognitive framework to recognise your own limitations. You cannot identify the gaps in your knowledge because you do not have sufficient understanding to know where they are. The result is a systematic overconfidence that is not a personality failing. It is an epistemic one, a structural feature of low-expertise states that produces confidence precisely in the absence of the information that would justify it.

Drivers rate themselves as above average, a statistical impossibility that illustrates how reliably most people overestimate their competence in familiar domains (Svenson, 1981 and Koppel 2023)

The tendency to believe we are better than average is one of the most robust findings in social psychology. In a classic study, Svenson (1981) asked participants to rate their driving safety and skill relative to others in the study. The majority rated themselves as above the median, even though it is statistically impossible for most people to be above average. This data has been replicated. In a large, preregistered replication and extension by Koppel et al. (2023), the researchers reported that "the majority of participants rated their driving skill and safety as above average," with results remaining stable across several response scales. Remarkably, over 90 per cent of participants believed they were safer than the median driver. Real-world data tell a different story. A 2021 RAC survey found that 73 per cent of young Australian drivers rated their driving skills as better or much better than their peers, despite 40 per cent of that same group reporting involvement in a crash. The better-than-average effect illustrates how the brain protects self-image by selectively interpreting evidence in our favour. This serves to remind us that confidence is not always calibration, and that without deliberate reflection, we often overestimate our competence while underestimating risk.

The mirror image of Dunning-Kruger is equally significant and less often discussed. People with genuine expertise in a domain consistently underestimate the difficulty others will have in acquiring it. They have forgotten what it felt like not to know, and they underestimate the cognitive distance between their current understanding and that of a beginner. This is the curse of knowledge, a complementary blind spot in which depth of expertise produces a systematic failure of empathy for the less expert, making teaching, communicating, and leading across expertise levels substantially harder than they need to be.

Together, these two patterns describe a landscape in which the people most confident about a domain are often those with the least grounds for confidence, while the people most qualified to have confident views often underestimate how much of their understanding others do not share. Navigating this landscape well, as a professional, a leader, or anyone trying to learn something genuinely new, requires a specific kind of skillset: the capacity to hold uncertainty without distress, to distinguish between what you know and what you think you know, and to remain genuinely open to evidence that contradicts the working model you have built.

This is a skill, and like all skills, it can be developed with the right understanding, the right practice, and, crucially, the right relationship to the discomfort that genuine intellectual updating produces.

The Narratives That Protect Us

What are the specific stories that surface when professional identity encounters an unfamiliar challenge?

When I work alongside senior professionals managing the transition, I hear variations of the same handful of narratives that suggest they are not individual responses but cultural ones, shared scripts that the professional brain reaches for when its sense of expertise and identity is under pressure.

'I am not technical.' This account places the barrier in a fixed personal attribute rather than a learnable skill. It converts what is a learning challenge into an identity statement, and identity statements are far more resistant to change than capability gaps, because changing them does not require acquiring new knowledge and a self-conception that may have been in place for decades.

'I am too established to start over.' This account explains the relationship between experience and learning. In a rapidly changing environment, established expertise is most valuable when combined with a willingness to extend and update it. But this viewpoint treats seniority as a reason to stop learning, precisely at the moment when senior professionals' learning matters most, both to their own continued contribution and to the cultures they lead.

'This is just hype.' This accounts for scepticism as a substitute for examination. Healthy scepticism is one of the most valuable cognitive tools available, but it requires actually examining the evidence before forming a judgement. The 'just hype' story frequently serves as a preemptive dismissal: a way of resolving the discomfort of uncertainty by closing off the inquiry that would reduce it. It protects the existing worldview at the cost of the information that might improve it.

Those stories are like conclusions. They are stories. The difference matters because conclusions are things we accept, while stories are things we can examine, and sometimes, with enough honesty, revise.

What all three narratives share is a function: they protect something. They protect the existing sense of professional identity, the investment already made in a particular set of skills, and the comfort of a worldview in which current expertise is sufficient for future demands. This is entirely understandable, and managing the challenging professional environment is increasingly costly.

The invitation is to notice these accounts as they arise. Not to dismiss them, because they usually contain something real, and not to accept them as fixed truths, because they usually contain something that is rationalisation rather than reasoning. But to hold them up to the light and ask: is this a conclusion I have reached by examining the evidence, or is this a story I am telling to avoid having to?

AI as a Cognitive Mirror

Artificial intelligence does something that few other professional developments have done as directly or as quickly: it holds up a mirror to the quality of our thinking.

When a generative AI tool produces a clearer formulation of a problem than we had reached ourselves, when it surfaces a body of research we had not found, when it identifies a logical gap in an argument we considered complete, these are not simply demonstrations of computational capacity. They are invitations to examine what we thought we already understood.

The two possible responses to this experience are ones most professionals will recognise from their own reactions. The first is defensiveness, a move to protect the existing sense of expertise by discounting the AI's output, emphasising its errors, or framing the comparison as unfair. The second is curiosity, a willing desire to engage the AI, to ask what it suggests about the limits of the existing understanding, and to use the encounter as data about where general learning is available.

The response a person makes to these instances is not random. It is largely predicted by where their sense of professional identity is anchored. A professional whose identity is organised around being the expert, the person who has the answers, who is looked to for certainty, whose value is seen through their knowledge, will experience AI's capability as a threat to that identity and will tend toward defensive responses. A professional whose identity is organised around being an open-minded and rigorous thinker will experience AI as an opportunity and will tend toward curious responses.

The real disruption that AI produces is not technological. It is psychological because it accelerates the rate at which our gaps become visible and forces us to question what our relationship to our own limitations actually is.

This is the sense in which AI functions as a cognitive mirror. Not because it shows us a perfect reflection of our thinking, but because AI has its own hallucinations, systematic errors and limitations. Because the encounter with an AI tool that can replicate some of what we do, quickly and at scale, surfaces questions about what we are actually offering that the tool cannot. What are the uniquely human aspects of expertise, where judgement and context and social intelligence and ethical reasoning are genuinely irreplaceable, and where what we had taken to be expertise was actually familiarity, the comfort of a well-worn path rather than the depth of genuine understanding?

Those are not comfortable questions. They are, in my experience, among the most confronting ones available.

The Four Stages of Competence and the Hardest Transition

In 1969, Martin Broadwell described a model of learning that has since become one of the most practically useful frameworks in professional development. He called it the four stages of competence, and it describes a progression that anyone who has learned anything recognises from the inside. Sidenote here, the originator of this process became a topic of debate on Wikipedia.

The first stage is unconscious incompetence: you do not know what you do not know. You may be performing poorly in a domain without realising it because you lack the metacognitive framework to accurately assess your own performance. This is the Dunning-Kruger territory, confident precisely because uninformed.

The second stage is conscious incompetence: you become aware of the gap between your current understanding and the understanding the domain requires. This is often triggered by an external event, a piece of feedback, a comparison, an encounter with expertise that exceeds your own, or an AI tool that surfaces a gap you had not noticed. It is uncomfortable because it involves trading certainty for awareness, and, as we have discussed, the brain finds certainty more comfortable than awareness.

The third stage is conscious competence: you have acquired the skill or knowledge, but applying it still requires deliberate effort and attention. You are performing well, but the performance is not yet automatic; it requires focus, concentration, and active management of a process that has not yet become second nature.

The fourth stage is unconscious competence: the skill has been sufficiently practised that it operates automatically. You perform well without having to think carefully about how you are performing. This is what expertise feels like from the inside: the smooth, effortless fluency that comes from deep, repeated practice.

The hardest part of real learning is moving from stage one to stage two, from not knowing what you don't know to realising that you don't know. It can feel like a loss, but it is actually the first step toward growth.

The reason this model is especially relevant in the AI era is that AI dramatically accelerates the transition from stage one to stage two. It surfaces gaps, exposes assumptions, and compresses the time between confident performance and being confronted with evidence of limitations. For professionals who have built careers on unconscious competence in deep domain expertise, the experience of AI revealing relevant knowledge or capabilities they were unaware of can feel not only uncomfortable but destabilising, and a challenge not only to their understanding but also to their professional identity.

Understanding that this experience is stage-two learning and that the discomfort of awareness is the entry point to genuine growth. The professional who can recognise the feeling of conscious incompetence as a signal of the right kind of movement, rather than a threat to be defended against, has access to a quality of continued development that the professional who remains in stage one, protected by certainty, does not.

Rethinking as a Learnable Skill

The framing I want to offer here and again, as an invitation rather than an instruction, is that the capacity to update one's beliefs is not a personality trait distributed unevenly across the population. It is a cognitive skill that can be practised, developed, and made more available under the conditions of pressure and uncertainty, where it is most needed.

The research on what psychologist Adam Grant calls 'rethinking, the deliberate review in light of new evidence, identifies several specific practices that support it. They are worth naming precisely, because the general instruction to 'be more open-minded' is both true and unhelpfully vague.

Separate identity from position

The single most powerful predictor of whether someone can update a belief under challenge is whether their sense of self is attached to the belief's content or to the quality of their reasoning process. A person who identifies as 'someone who holds position X' will defend X when challenged, because challenging X is challenging them. A person who identifies as 'someone who thinks carefully about difficult questions' can update X when the evidence warrants it, because doing so is consistent with, rather than threatening to, their self-conception.

At times of disagreement, a simple internal script can support this shift: "My worth is not attached to being right. Updating my view is a sign of skill, not a threat to my identity." Or "I am the kind of person who values accurate understanding more than defending a position." Practising this self-talk, even silently, anchors you to the idea that being open to new evidence is a strength, not a weakness. This short pause in self-talk can create the psychological safety needed for genuine rethinking.

It has a practical corollary: the most up-to-date professionals tend to be those who have practised describing their views as current working models rather than fixed positions. Not 'I believe X' but 'my current understanding is X, based on Y and Z.' It frames the belief as contingent on evidence rather than identity, which makes it easier to revise when better evidence arrives.

In practice: When you find yourself defending a position under challenge, pause and ask: Am I defending this because the evidence supports it, or because it is mine? The question is not comfortable. In my experience, it is almost always clarifying.

Treat discomfort as information, not as a signal to stop.

The discomfort of cognitive dissonance, the unsettled feeling of encountering information that challenges an existing belief is, as we have discussed, has a stressful signature that can be mapped in the brain circuits and produces the impulse to resolve the discomfort as quickly as possible.

The default position is dismissal of the challenging information. The growth-mindset-shifting approach is to stay with the discomfort long enough to examine it and ask what, specifically, is being challenged, whether the challenge is valid, and what updating would require. This is harder than dismissal and more valuable.

The practice of treating mental discomfort as a signal of potentially important information, rather than a threat to be neutralised, can be deliberately cultivated. It does not become comfortable with practice. It becomes more manageable, which is something different and more useful.

In practice: When you notice the urge to dismiss a challenging idea quickly, particularly one that arrives from an AI tool, a junior colleague, or someone outside your field, try staying with it for sixty seconds longer than feels comfortable. Ask one question: what would I need to believe for this to be right? You are not committing to an agreement. You are practising the mental flexibility that genuine updating requires.

Seek disconfirmation deliberately

Confirmation bias, or the tendency to seek out and remember information that corresponds with existing beliefs, is a feature of the information-processing system described at the beginning of this chapter. You cannot simply decide not to have it.

Deliberately seeking out the strongest version of the opposing argument, what philosophers call 'steelmanning', counteracts confirmation bias not by eliminating the tendency to seek confirming information but by supplementing it with a deliberate practice of seeking disconfirming information. Reading the most rigorous critique of your position and asking someone you respect, who disagrees with you, to explain their strongest objection. Using AI tools specifically to generate counterarguments to a view you hold.

None of this requires you to abandon your existing understanding. It requires you to test it, which is what scientific thinking asks of any hypothesis, and what genuinely expert judgement does automatically in mature practitioners.

In practice: For any significant belief or position you hold professionally, ask an AI tool to generate the most compelling counterargument available. Read it as if a trusted colleague had offered it. Then ask: Does this change anything? The answer may be no. But the practice of asking keeps the belief in the domain of examined positions rather than unexamined assumptions.

Belief, Biology, and What Remains Possible

I want to close this chapter where it began, with the relationship between belief and biology, and with a comment on the changes that follow when we take that relationship seriously.

The story I held about psychiatric medication was not simply a professional opinion; it influenced my research approach. Because the beliefs that shape professional recommendations shape the advice that forms choices that shape the lives that result from them. Belief, when held by someone with influence over others, is not merely personal; it has downstream consequences that extend well beyond the person who holds it.

This is true for every professional who leads, advises, or influences, which means almost every professional in every field. The beliefs we hold about what is possible, what is acceptable, what constitutes legitimate intervention and genuine expertise, shape the environments in which the people we work with make their choices and build their lives. When those beliefs are examined carefully and updated accurately, the consequences are beneficial. When they are held as unexamined certainties, the consequences are whatever our blind spots produce.

The invitation this chapter offers is not to doubt everything or to treat all beliefs as equally uncertain. It is something more specific: to hold your current understanding with the kind of confidence that distinguishes genuine expertise from identity-protective certainty. To know the difference between a conclusion you have reached by examining evidence and a story you are telling to avoid having to. To treat the discomfort of genuine challenge, not as a threat but as the entry point to the kind of learning that makes expertise worth having in the first place.

Intelligence is not knowing. It is knowing how to examine what you know, update it when the evidence warrants, and remain genuinely curious about the distance between your current understanding and what is actually true.

Almost every hand goes up in that room of senior professionals when I ask who considers themselves above average at critical thinking. I understand why I have raised my own hand in similar moments without pausing to examine whether it was justified.

The question is not whether we are capable of rigorous thought; the question is whether we are applying that rigour to the beliefs we are most invested in, including, and perhaps especially, the beliefs about our own thinking.

That, in my experience, is where the most important work is.

And it is never quite finished.

CHAPTER NINE

Why Are You Here?

Purpose, meaning, and the neuroscience of a life worth living

The question that organised my life arrived in 1989, in a psychiatric ward, watching my sister receive an injection that she had not been consulted about, in a system that did not fully understand what was happening to her brain. I was sitting in a room with her, helpless in the way that only people who are watching something terrible happen to someone they love to understand, and I asked not quite to anyone, not quite aloud in my mind. Why was this happening to her, and we must not know what is happening inside the brain?

I did not know, then, that the question would not be resolved. That it would instead expand and deepen and change shape over the following thirty-five years, pulling me through pharmacy and neuroscience and research and teaching and writing, always toward the same underlying thing: understanding what goes wrong in the brain, and what it takes to help it heal.

I didn't realise the question was my purpose.

What I have come to understand through the research, through the decades of work it produced, and through the specific experience of having lived inside a question long enough to see what it builds, is that purpose is not something you find, the way you find an object you have misplaced. It is something you follow: a thread that was present before you knew to look for it, that has been pulling you forward in various forms throughout your life, that you can identify in retrospect and then choose, with more intention, to follow forward.

Purpose is not what you do. It is why you do it and the through-line connecting your daily actions to something larger than the mere accumulation of those actions. It is the reason the work matters, even on the days when doing it is difficult.

What the Research Shows

People with a strong sense of purpose have, on average, a lower risk of cognitive decline and Alzheimer's disease, reduced cardiovascular risk, lower levels of chronic inflammatory markers, better sleep quality, and longer lifespan. The magnitude of these effects is not trivial. The 10-Item Purpose in Life Measure includes statements such as "I feel good when I think of what I've done in the past and what I hope to do in the future"; "I have a sense of direction and purpose in life"; "I enjoy making plans for the future and working them to a reality" and "I am an active person in carrying out the plans I set for myself," alongside reverse-scored items such as "My daily activities often seem trivial and unimportant to me" and "I sometimes feel as if I've done all there is to do in life."

In large longitudinal studies, Patricia A. Boyle, PhD, and her colleagues from the Rush Alzheimer's Disease Centre evaluated more than 1,200 older adults without dementia. Researchers followed people over several years to understand what predicts longer life. At the start, participants were asked about their sense of purpose and how much they felt their lives had meaning and direction. Over the next five years, those with a stronger sense of purpose were significantly less likely to die than those with lower scores. This effect was not small. People with a high sense of purpose had a markedly lower risk of mortality, even after accounting for age, health conditions, depression, income, and other factors.

The finding was consistent across men and women, and across different backgrounds.

What this tells us is simple but profound. Purpose is not just a feeling. It is a biological and psychological driver of health. It shapes how we live, how we cope, and ultimately, how long we live.

Purpose also changes how the brain processes stress. When effort is directed toward something meaningful, the same physiological stress response, the cortisol, the amygdala activation, the sympathetic arousal, is processed through a different cognitive frame. Challenge rather than threat, and the distinction matters not only psychologically but also physiologically: challenge responses produce less sustained inflammatory activation than threat responses, meaning that purposeful engagement with difficult work is genuinely less physiologically costly than effortful engagement with work that feels meaningless.

Ikigai

The longest-lived populations on earth are not the ones with the most sophisticated healthcare, the lowest stress, or the most favourable genetics. They are the populations in which people maintain a strong sense of purpose well into old age, in which there is no sharp boundary between working life and purposeless waiting, and in which the question of what one is for remains alive and actively answered into the ninth and tenth decades.

The Japanese concept of ikigai, loosely translatable as 'reason for being,' the thing that gets you out of bed in the morning, has attracted attention from longevity researchers because it appears repeatedly in the profiles of the unusually long-lived populations of Okinawa. The very elderly in Okinawa, whose ikigai has been documented, tend gardens, make traditional crafts and care for grandchildren and great-grandchildren.

They do not retire in the sense that most Western cultures understand retirement, as the end of making a valuable contribution and the beginning of deserved leisure. Their purpose evolves as their circumstances change, but it does not disappear at an arbitrary age defined by actuarial tables.

The contrast with cultures where retirement does mean ending work, where older people are subtly or explicitly positioned as having nothing left to contribute, where the loss of professional identity is not replaced by other sources of meaningful interaction, is visible in the cognitive and health outcomes that follow. The acceleration of cognitive decline after retirement is well documented in populations where retirement leads to a genuine loss of purpose and is largely absent in populations where it does not.

Purpose Is Not Ambition

One of the most common misunderstandings of purpose, in my experience of talking about this with professionals across many fields, is the idea that purpose is achievements, with grand mission statements or bucket lists or the kind of significance that makes the biography shelves.

I held this misunderstanding myself. In the years when I was defining my purpose as understanding and solving mental illness, as making the breakthrough discovery that would alter everything, it was not without purpose, but I was holding it in a way that made it fragile. Because the gap between the actual work of science, which is incremental, uncertain, and frequently unglamorous, and the story I was telling about what the work was for was large enough that ordinary difficult days appeared as failures rather than as ordinary difficult days.

What I came to understand, partly through the research and the experience of sitting in Berkeley at the peak of the achievement I had been pursuing and finding it insufficient, is that purpose is most durable when it is most specific about the why and the what. The through line that has organised my professional life, understanding suffering in order to reduce it, is not tied to any particular method of understanding or any particular domain of suffering. It has expressed itself as pharmacy, neuroscience research, teaching, writing, public engagement, and this book. The core is stable; however, its expression has evolved continuously and will continue to do so.

Purpose doesn't have to be grand to be real. A gardener who tends a community plot, a parent who shows up fully for their children, a colleague who consistently makes the team around them more capable, these are purposeful lives. The scale is not the variable that matters.

What matters is that the through-line is genuinely yours, not inherited from what your profession expects, or what your family assumed, or what ambition dictated before you had enough information to know what you actually valued. And that the thread is alive: actively expressed in what you do, not simply as a belief about what you are supposed to be for.

When Purpose Gets Lost

I have watched purpose disappear from people's lives, and the pattern is consistent enough to name.

It happens most often at transitions: retirement, the departure of children, the end of a relationship, the loss of a role that had been providing meaning without the person necessarily recognising it. These transitions are losses not only of the specific thing that has ended but of the frame through which the days were organised and the effort felt justified. Without that frame, the ordinary activities of a life continue, meals, sleep, errands, appointments, but the sense that they are building toward something, that they are expressions of something worth expressing, quietly drains away. And the person, from the inside, experiences something that is difficult to name, not depression exactly, not grief exactly, but a kind of hollowness that sits beneath the ordinary days.

The question in these moments is not what my purpose was in the past tense; it is asking what my purpose is now, given who I am and what my current circumstances make possible. The through line does not disappear when a role ends; it needs a new expression.

In practice: If you are undergoing a transition in which the structure that previously carried your sense of purpose changed or ended, try this: name the through-line rather than the expression. Not 'I was a teacher' but 'I help people understand things that matter to them.' Not 'I was a manager' but 'I create conditions in which people can do their best work.' The through line is almost certainly still available to you. The question is: what form does it take now?

Finding the Thread

If you are uncertain about your purpose, if it feels like one of those questions that other people seem to have answered more clearly than you have, I want to offer a different frame than the one usually applied to this question.

Most guidance on finding purpose directs people forward: toward goals, aspirations, the future self they want to become. This can be useful, but it tends to produce purposes that are more about ambition than about genuine identity, about what seems impressive or worthy rather than what is alive in the person doing the thinking.

As Steve Jobs said in his 2005 Stanford commencement address:

"You can't connect the dots looking forward; you can only connect them looking backwards. So, you have to trust that the dots will somehow connect in your future."

Looking backward is often more revealing. The thread of your purpose has almost certainly been present throughout your life, expressing itself in various forms that you may not have recognised as connected. What questions have kept pulling at you, even when you were not pursuing them professionally?

What problems make you genuinely angry in a way that signals something real about what you think matters? What are the moments, across your life, when you felt most fully yourself, most alive, clear and exactly where you were supposed to be?

The thread is in those answers. It is usually simpler than the grand purpose narratives suggest, and more durable: regarding connection, understanding, beauty, justice, building, protecting, teaching, or healing. Some version of one of those ancient human motivations, expressed in the specific way that your particular history and capacities make possible.

And when you lose sight of it, which happens, because the demands and urgencies of ordinary life are good at obscuring it, use the looking-backward question to find it again.

In practice: Write down three moments in your life when you felt most fully yourself, most engaged, purposeful, or exactly where you were meant to be. They do not need to be career moments; they can be anything. Then look for what they have in common. What is the shared element that is almost always a version of your through-line? It has been there longer than you realised.

The Question That Keeps Going

My purpose started with a question I asked in a psychiatric ward corridor in 1989. It has produced decades of work, a research career, a podcast, and many books. It has changed the specific form of its expression many times. What Francesca's story gave me was not a mission to complete but a direction and a reason to keep moving forward, to keep asking, and to keep bringing whatever I currently understand to the people who might use it. That direction is alive in every chapter of this book and in the conversations I have with people who are struggling with their own minds or the minds of people they love. It is alive in the research I still read and the questions I still pursue. It is, in the most practical sense of the word, why I am here.

What is your thread? Not your job title or your achievements. Not the role you are most identified with in the eyes of others. The thread that has been pulling you forward since before you had words for it. And the people around you, the ones who would be shaped by what you are for, the ones who need what your thread has to offer, they need it too.

CHAPTER TEN

Stepping Up

Building intelligence worthy of the next generation

I was sitting in the Concert Hall of the Sydney Opera House when Tim Berners-Lee said something that has not left me since.

He was speaking about the early decisions around the web, the moment when he and his colleagues chose to make its core protocols open, such as HTTP, HTML and URLs. They could have been patented and controlled. Berners-Lee could have become extraordinarily wealthy, and the architecture of the web would have been determined by a different set of interests than those that shaped it; he offered the protocols freely. And that choice made by a small group of people who could not have fully anticipated its consequences determined, more than perhaps any other single decision, who could build on the web, who could participate in it, and who could benefit from it.

As I listened, I was thinking about my own family. That same week, my daughter had begun medical school, and my son was working on building AI systems. One is a step into a profession grounded in ethics and human care, and the other is contributing to the technology that will reshape how that care is delivered.

And what I kept returning to, listening to Berners-Lee describe the consequences of a decision made before the web was yet a word most people knew, was this: the choices being made right now about AI, transparency, accountability, governance, and who benefits will shape the professional environment my children inherit. The design decisions being embedded in AI systems today will be as consequential as the decision to make HTTP a public good. And most of those decisions are being made in rooms where most professionals are not.

The question is not whether AI will transform your field. It already is. The question is whether you will help define the standards by which it operates or whether you will discover, after the fact, that someone else defined them for you.

Watching From the Inside

As someone connected to researchers actively developing AI systems, I have watched this transition from an unusual vantage point close enough to the technology to understand what it can do, and close enough to the institutions it is entering to see what it is doing to them.

The debates inside universities are practical and immediate in ways that the broader public conversation often is not. What constitutes authorship when a system generates the first draft? How do we assess learning when synthesis and summarisation can be automated? What does original contribution mean when the most sophisticated pattern-matching tool in history is available to every student with an internet connection? How do we maintain professional standards, the standards that make a degree mean something, that make a qualification trustworthy, without pretending the technology does not exist?

These are not abstract questions; they are questions being answered, institution by institution, profession by profession, right now, often under time pressure, without adequate frameworks, and without the input of the people whose expertise should be relevant to answering them well.

The people building AI systems are not, for the most part, careless. The researchers I know are asking serious and sustained questions about alignment, bias, safety, and misuse. But there is a category error that keeps appearing in the public conversation about AI governance, and I want to name it directly: engineers cannot define medical ethics. Model developers cannot determine legal obligations, and data scientists cannot set educational integrity frameworks. Those responsibilities belong to professions and to the accumulated expertise, ethical traditions, and institutional accountability that professional practice represents.

Ethical AI is not a technical add-on that can be appended to a system after the architecture is set. It is a professional obligation that must be present in the design process, deployment decisions, governance structures, and ongoing monitoring. And it requires professionals, not only technologists, to be actively present in those processes.

The Governance Gap

The voices calling for human monitoring of AI have been consistent and credible. Stuart Russell, one of the creators of AI, has argued that the main challenge is to ensure it remains beneficial rather than merely capable. Timnit Gebru has carefully documented that without accountability mechanisms, bias and inequality do not merely exist in AI systems but also scale with them. Margrethe Vestager has insisted that humans must retain meaningful control over high-stakes automated decisions. Tim Cook has reminded the technology industry that the purpose of technology is to serve humanity, a principle that sounds obvious until you observe how often deployment decisions are made based on what is technically possible and economically attractive rather than what is humanly beneficial.

They are governance requirements in a democratic society and are ultimately the responsibility of the professions and institutions that a society trusts to define and sustain standards. The phrase "human-in-the-loop" has become common in AI governance discussions, and it is worth being precise about what it must mean. It cannot mean nominal approval, a human technically signing off on a decision that an algorithm has effectively already made, with no genuine capacity to interrogate or override it. It must mean identifiable human responsibility in high-risk contexts: a named person with the expertise and authority to examine the basis of an AI recommendation and the accountability to answer for the results. Responsibility cannot be delegated to code; this is not a technical limitation but a principled one.

What does real professional accountability look like in practice? To move these principles into action, each profession that uses AI for consequential decisions should define clear requirements for "named human responsibility." Here are a few checklists to make this concrete:

For lawyers:

— A designated lawyer must review all AI-generated recommendations, draft documents, or analyses before client delivery or court submission.

— The lawyer must be able to document not only that the output was reviewed but also what specific legal judgements were made and points of disagreement or correction.

— The responsible lawyer's name and contact must be attached to any AI-assisted output, ensuring traceability and accountability to the client and the courts.

For doctors:

— A named clinician must review, interpret, and, where needed, override any AI-generated diagnosis, triage suggestion, or treatment plan before action is taken.

— The clinician must record in the patient file which parts of the AI's output were accepted or rejected and the rationale for each.

— The responsible medical professional must be identifiable in all records, so accountability is clear to both the patient and any future reviewers.

For teachers or educators:

— An assigned teacher must validate all AI-generated feedback, grades, or recommendations that affect student records or learning pathways.

— The teacher should communicate to students and parents when and how AI was used, and provide an avenue for appeal or explanation

— The teacher's name and professional role should be linked to any AI-assisted assessment, ensuring human monitoring and opportunity for correction.

The specifics will vary by field, but the underlying pattern must be the same: professional accountability is visible, auditable, and personal. It is not sufficient for a system to "pass through" human hands; those hands must clearly grasp, assess, and be answerable for the consequences.

The Economic Dimension

There is a dimension of the AI transition that professional communities have been slower to engage with than the ethical and governance questions, and that I think deserves more direct attention: the economic one. There is no question that artificial intelligence increases productivity. The real issue, and one that matters to professionals in every field, is how those gains are shared and who ultimately benefits from them.

The pattern that has characterised previous waves of automation helps to understand the consequences. When productivity gains are captured primarily by those who control the technology, data, computing infrastructure, and deployment platforms, while the labour market moves slowly as capital accumulates, inequality widens. Not because anyone necessarily intends this outcome, but because it is the default outcome when distribution is left to market factors, without the countervailing pressure of professional standards, regulatory structures, and institutional governance.

AI can compress roles in analysis, documentation, and administration across law, medicine, finance, and education. It can also create new roles, expand access to expertise that was previously scarce and expensive, and reduce the burden of the administrative and procedural tasks that have crowded out the genuinely human work in many professional fields. Which of these outcomes predominates in any given profession will depend, to a significant degree, on the choices that professionals in those fields make about how AI is adopted, what roles it is permitted to occupy, and what governance structures are put in place to protect the people whose work it affects.

If professionals disengage from these conversations, the decisions default to market forces. And market forces, unguided, are not designed to ask who benefits, who bears the risk, and who is left behind.

These are not questions that technology companies are best positioned to answer. They are questions that require the domain expertise, the ethical traditions, and the institutional accountability that professions represent. Crucially, shaping fair and effective AI systems also requires involving the people most affected by those systems. Inviting students, patients, clients, or other community members into the design and decision-making process helps ensure that AI tools are genuinely responsive to varied needs and culturally sensitive in their outcomes. Participatory design transforms AI adoption from a top-down initiative into a shared stewardship project. Which means they require professionals to be present in the rooms where the answers are being decided, not after the fact, as critics, but in the process, as participants, and as co-designers alongside those whose experiences the technology will shape.

What Stepping Up Actually Means

People are helping build AI systems, and others are entering professions that will use them. They represent the two sides of the relationship that the AI transition is creating in every field: the people building the technology and the people whose practice it will reshape. The standards established now will shape how they work, how they are judged, and how they are trusted by their patients, their clients, their institutions, and ultimately by the public.

If you are wondering what it actually looks like to step into the role of steward for AI in your profession, start with a single concrete action. Join an AI ethics or governance working group in your professional association or volunteer for your institution's technology standards committee. If no such group exists, ask your leadership to form one, or propose that your organisation participate in wider industry collaborations to set best practices for AI adoption. Even a single conversation or contribution can put you at the table where decisions are taking shape.

For professionals just beginning to learn about AI governance, there are other accessible first steps. You might attend a webinar or public panel on AI ethics in your field or subscribe to the newsletter of a reputable AI and society research group. Reading a key policy report, such as your professional association's guidelines on AI or a recent government white paper on AI regulation, can also provide useful grounding. Starting with these kinds of low-barrier actions is often enough to move from hesitation to meaningful engagement. Taking a first step transforms concern into agency and indicates that shaping the future is part of your responsibility. The stewardship the next generation needs begins with professionals willing to act, however small the move may seem at first.

Stepping up is not a dramatic act and does not require becoming a technologist, an AI policy expert, or a public intellectual. It requires something more modest and durable: engaging with the questions the AI transition raises in your own professional domain, with the seriousness and sustained attention those questions deserve.

It means learning enough about how AI tools are entering your field work so you can evaluate their outputs critically rather than accepting them uncritically. It means participating in the conversations within your institution, your professional association, and your regulatory body about the standards that should govern AI use in your domain. It means mentoring the next generation in judgement, ethical reasoning, and social intelligence that remain irreplaceable regardless of how sophisticated the tools become.

It means being willing to say, when an AI system produces a recommendation that your professional judgement tells you is wrong, that your judgement takes precedence and being able to articulate why.

And it means understanding that the choices being made now, in the early years of AI's integration into professional practice, have the same structural significance as the choice Tim Berners-Lee made about open protocols. They will determine, for a generation, who benefits, who participates, and what the technology serves.

When machines increase our power, we must increase our wisdom. This is not about resisting progress. It is about shaping it deliberately while there is still meaningful room to do so.

We cannot later claim that we did not see the implications. We are in the room while these systems are being designed, funded, deployed, and normalised. The question is what we do while we are there. The future will be intelligent, and the people who shape it need to be wiser.

CHAPTER ELEVEN

Our Thriving Minds

The work now becomes personal

We have covered a great deal of ground together. From a psychiatric ward in 1989 to the neuroscience of neuroplasticity. From dopamine loops and digital distraction to the governance of artificial intelligence. From the biological fundamentals of brain health to the cognitive traps that keep even intelligent people from updating their most important beliefs. From the distracted attention of modern professional life to the responsibilities that the AI transition places on every field.

We have examined how systems shape brains and how brains, in turn, shape systems.

Now the lens turns inward. Because knowledge, by itself, does not create change. You can understand neuroplasticity intellectually and still repeat the same habits. You can agree with every principle in this book and still hand your attention over to the nearest notification every morning. You can recognise the value of deep work, of regulated stress, of sleep and movement and genuine connection, and still choose, in each particular moment, the easier thing. The question is no longer what should be done. You know that, or you know more of it than you did. The question is whether you will do it.

Progress, Not Perfection

I still reach for chocolate when I am stressed, sometimes before I have consciously registered it. My brain still fires at things that do not warrant the response it produces. I still catch myself at the end of a day that was full of activity, with the unsettling sense that I did not think deeply about anything that mattered. I still open my phone in moments of mild boredom or discomfort, and I know, from everything discussed in this book, that this serves neither my attention nor my mood.

The difference between now and the person I was in the years I have described in these pages is not that the patterns have vanished. It is that I notice them and have enough self-awareness to recognise when I am in a pattern rather than in a genuine response. And I have enough experience of having changed patterns I thought were fixed to believe, not merely intellectually but from evidence gathered in my own life, that change is possible.

That is what a thriving mind looks like: not the absence of struggle, but the capacity to work with it. Not the elimination of stress, but the ability to regulate it. Not perfect habits, but enough consistent habits to tip the balance to ensure that the days on which the brain gets what it needs outnumber the days on which it does not, by a margin that is slowly, cumulatively, moving in the right direction.

Improvement over perfection is how neuroplasticity works. The brain changes through repetition, not because you decided once and for all to be different.

Neuroplasticity responds to practice, and the brain that changes is the one repeatedly and consistently given a different experience to wire itself around.

What Changed for Me

When I look back at my life in Berkeley, running a research lab and managing what felt like a thousand moving parts, I can see something clearly now that I could not see then. I was holding a lot together, spinning multiple plates at once, and for a long time, I believed I was doing it well. I thought I was being efficient, capable, even clever in how I handled the pressure. But what I could not see was the cost. Beneath the surface, I was under constant stress, and my ability to think deeply was slowly eroding. Eventually, that way of working caught up with me.

What has changed is not one thing. It is a set of shifts that, over time, have changed how I live, think, and feel. My relationship with food is one example. For years, it followed a familiar pattern of restriction, indulgence, and guilt. That cycle is no longer part of my daily life. I did not change this through willpower. I changed it by understanding what was driving it. Once I understood the mechanism, something subtle but important happened. A pause appeared between the impulse and the action. At first, that pause was small and inconsistent, but over time, it became enough to choose differently. That is how change happened.

It was not linear. There were many times when old patterns returned, especially under stress. What changed was how I responded. Instead of seeing those moments as failure, I began to see them as part of the process. Each time I returned to a new pattern, I was strengthening it. This is what neuroplasticity looks like in real life. It is not about getting it right every time. It is about returning, again and again. Movement is now part of my life in a different way. I do not exercise because I have suddenly become more disciplined. I move because I have repeatedly felt the difference it makes to my thinking, my mood, and my capacity to cope. It has shifted from something I should do to something I rely on. My relationships have changed, too. I have fewer of them, but they are deeper and more meaningful. I used to think that more connections meant less loneliness. What I have learned is that the quality of connection matters. Depth creates something that a surface-level connection never could. My ability to focus has also returned, in a way that feels stronger than it was earlier in my life. I now understand what deep thinking requires, and I protect it. With that has come something I did not realise I had lost for a time: genuine intellectual enjoyment in my work.

None of this happened quickly. There was no single turning point. These changes came from small, consistent actions, repeated over time. On their own, they were barely noticeable. Together, they have changed everything.

What Is Possible for You

Your brain is more capable of change than you probably believe, at whatever age you are reading this. The research on neuroplasticity does not set an upper age limit. It does not conclude that the brain's capacity for adaptation is exhausted. What it shows is that the capacity for new learning, for the formation of new patterns, for the gradual rewiring that practice produces, remains available across the lifespan.

What changes as we age is not the brain's willingness to change. It is the conditions under which change happens most readily, and the timescale over which it becomes visible. An older brain, on average, requires more repetition to consolidate new learning, more recovery time between demanding cognitive efforts, and more deliberate attention to biological conditions, such as sleep, movement, nutrition, and regulated stress, that support mental health.

Start small by choosing one practice and sustaining it long enough to see its effect. Then add another and build slowly, trust the process, and give your brain the time it needs to wire the new experience into something durable. One effective way to support this is by tracking your habits as you build them. Consider a daily check-in that takes no more than 1 minute: each morning or evening, mark a tick or cross in a physical journal, use a simple checkbox in a note's app, or try a habit-tracking app that sends gentle reminders and lets you log your progress with just a tap. Even this small, fast check-in can make your changes visible and help keep you motivated without adding extra burden to your day.

To help you notice meaningful changes and stay motivated, consider using simple self-reflection tools. One practical method is to keep a weekly self-check template that makes reflection quick and repeatable. Each week, set aside a few minutes, perhaps at the end of Sunday or the start of Monday, to note your responses to a few questions. Here is a sample you can adapt for your needs:

Weekly Self-Reflection Tools:

Sleep: On most nights this week, was my sleep restful or restless? (1 = very restless, 5 = very restful)

Focus: Did I find it easier or harder to concentrate on important tasks? (1 = very difficult, 5 = very easy)

Mood: What was my overall mood most days? (1 = very negative or anxious, 5 = very positive or calm)

Energy: Did I notice changes in my energy levels? Was I able to recover quickly from stress? (1 = low energy, slow recovery; 5 = high energy, quick recovery)

Notable Wins or Challenges: Briefly describe any 'small wins' or ongoing challenges you noticed this week.

You can print out a simple table or keep these questions in a notebook or digital document, checking off or rating each area on a scale of 1 to 5. These quick self-checks make progress tangible, highlight what is working, and offer early signals of the positive shifts you are making.

At the end of each week, spend a few minutes reflecting on your progress. Did you notice any shifts in how you feel or think? This kind of tracking creates accountability, makes progress visible, and helps you recognise the accumulation of small efforts over time. A year from now, you could be genuinely different, not because you resolved to transform yourself, but because you did small things consistently, and small things done consistently compound.

The Practices That Matter Most

If I had to reduce this entire book to the practices that matter most, the ones that research consistently supports, and that my own life consistently confirms, they would be simple. Resilience is built in small, repeated decisions, especially when the world looks unstable. The first decision happens within seconds of waking.

What happens in the first hour of the day? Do you swipe, squint, scroll, or do you stretch, stand, and scan the sky? The way you begin the day shapes the rest: silence your phone, stash it out of reach, stand by the window, and let your gaze widen. Start your day with action, not reaction.

When you deliberately widen your visual field, you recruit broader attentional networks, including the dorsal attention system and reduce stress. Expanding your gaze signals safety to the nervous system. In simple terms, you move from scanning for danger to perceiving joy. And that shift reduces stress.

Now add three specific things you are grateful for. Choose something concrete and real. The conversation that shifted your thinking yesterday. The fact that your body carried you through another day. The light in the room this morning. Gratitude changes what the brain is primed to notice. The brain is constantly generating predictions and then searching for evidence to confirm them. If you begin the day in a state of urgency and stress, your attentional systems will unconsciously seek out more of the same. If you begin with steadiness and a few moments of joy, the brain becomes more likely to detect opportunities, cooperation, and possibility.

Importantly, the brain is biased toward stress over pleasure. From an evolutionary standpoint, detecting threat was more urgent than appreciating safety. If you are not intentional, your attention will drift toward what is uncertain or wrong.

Being deliberate about the first inputs of the day is a simple way to retrain that bias.

The second practice is movement. Not the perfect workout but ten minutes of consistent movement is more powerful than the hour you planned but did not complete. Even brief daily movement increases cerebral blood flow and initiates the exerkine cascade that supports neuroplasticity. Over time, this consistency supports the brain's structural resilience.

In periods of disruption, when stress hormones run higher and cognitive load increases, daily movement becomes a stabiliser that quietly shapes the architecture of your brain. In an age of artificial intelligence, economic uncertainty, and rapid cultural change, thriving will not depend only on how fast you react. It will depend on how well you regulate.

Start your day before looking at a screen.

Set your mind to look for possibilities instead of threats.

Move your body daily, even for a brief time.

Resilience doesn't come from big gestures.

It's built in the first hour of your day.

Use your breath to regulate your state. Four slow, deliberate exhalations, longer out than in, calm the nervous system in seconds, not as alternative medicine but awareness of your agency over your physiology.

Protect your attention, and not by withdrawing from the tools that make modern professional life possible, but deliberately, by making the use of those tools a choice rather than a reflex, by protecting specific windows of undivided focus, by designing the environment so that the better choice is structurally easier than the worse one.

Choose real connection, as we are a social species that is regulated more effectively in the presence of trusted others than in isolation. Face-to-face connection, even if brief or imperfect, provides the co-regulatory input that online interaction cannot replicate.

Work with anxiety rather than against it, think of anxiety as a signal. When you notice it, ask what it is telling you and what threat it has detected, whether that threat is real and proportionate, and what action, if any, it warrants. Extract the information and take the appropriate action. Then practise letting it go, rather than holding the signal open after the message has been received.

Trust that your brain can change. This isn't just a motivational phrase; it's backed by neuroscience. Believing change is possible helps you act in ways that make change happen. What you believe can shape your biology.

Try to anchor your life to purpose. The centenarian research discussed consistently identifies future orientation, the sense that there is meaningful work still ahead, as one of the most powerful predictors of cognitive longevity and sustained wellbeing.

These practices are simple, repeatable, and repeatable is how brains change. The practice you will actually do consistently is worth more than the practice that is theoretically optimal.

For Francesca

I want to return, at the end of this book, to where it began.

My sister Francesca died in 2006, at thirty-seven, having lived the second half of her life inside a system that did not try to understand her brain and did not have the tools to help it. The treatments available to her were not designed for what she was experiencing. The environments she moved through, psychiatric wards, institutional care, and the margins of a professional life she had once aspired to, were not designed for healing. She was diagnosed with what was understood, in 1989, as schizophreniform disorder, and treated accordingly. Whether that diagnosis was accurate and what a more effective intervention might have looked like are questions I have spent 30 years thinking about without fully resolving.

What I understand now, but could not then, is that what happened to Francesca was biology meeting circumstance at one of the most vulnerable moments in a human life. The brain she was born with, shaped by the developmental environment of her earliest years, encountered circumstances that overwhelmed its capacity for regulation. The treatments she received addressed the symptoms as available at the time. They did not address the underlying early life experiences and trauma. The system and her family, doing their best with what they had, failed her, not through malice but through the limits of what was known.

We know more now, not everything and not enough, but substantially more than was known when Francesca first walked into a psychiatric ward, and the door closed behind her.

This book is part of what I can do with what I know. It is my attempt to make visible, to the professionals, the leaders, the parents, and the people navigating their own struggling minds who have read these pages, the connection between neuroscience and lived experience that might, in some cases, create earlier understanding, intervention, and different outcomes than the ones Francesca experienced.

Her life mattered, and her suffering shaped mine, and through mine, has shaped this book. It is not what I would have chosen for her. But it is what I can offer, in her name, to whoever reads it.

Science, Spirituality, and the Future of Neuroscience

For a long time, science and spirituality have been placed on opposite sides of the same question. Science has focused on what can be measured, tested, and replicated. Spiritual traditions have focused on meaning, connection, and experiences that are often deeply felt but harder to quantify. The divide has shaped how we think, how we practise, and how we understand the mind.

What is changing now is not the need for rigour, but the scope of what we can observe. Neuroscience is beginning to catch up with experiences that were once considered beyond its reach. We can now measure patterns of brain activity during meditation, track changes in neural networks during therapy, and observe how connections between people shape physiology in real time. We are starting to see that many experiences described in spiritual language, such as awe, presence, compassion, and connection, are not abstract ideas. They have biological signatures and they change the brain.

This does not mean that everything should be reduced to a scan or a signal. It means that we are expanding the boundary of what can be studied, and with that comes a responsibility to remain grounded in evidence while staying open to what we do not yet fully understand. The future of neuroscience will not be about choosing one perspective over the other. It will be about integration. It will involve bringing together what can be measured with what can be experienced, without losing the integrity of either. This is where the idea of thriving becomes important.

Thriving is not simply the absence of illness. It is the presence of capacity. It is the ability to think clearly, regulate emotion, connect with others, and find meaning in what we do. These capacities are shaped by the brain, but also by the environments we live in, the relationships we build, and the experiences we seek out.

In the years ahead, neuroscience will move beyond studying isolated brains and begin to focus more on brains in context. It will look at how we influence each other, how environments shape attention and behaviour, and how systems either support or undermine mental health. It will also need to address the realities of a digital world, where attention is constantly contested, and the pace of change is faster than our biology evolved to handle. We need approaches that people can use in their daily lives. Practices that support attention, movement, connection, and recovery. Ways of working and living that align with how the brain actually functions, rather than constantly pushing against it.

This is the direction of Thriving Minds. It is grounded in science, but it is not confined by it. It recognises that data matters, but so does lived experience. It focuses on what can be measured, while also paying attention to what can be felt. It brings together neuroscience, behaviour, and the environments we create around us. The future will not be defined by technology alone. It will be defined by how well we understand ourselves, how we care for each other, and whether we choose to build systems that support human capacity rather than deplete it.

Your Next Step

Close this book, and before you pick up your phone, or turn on the television, or return to the list of things waiting for your attention, take four slow breaths. Then name three things you are genuinely grateful for. Not three categories, not 'family, health, work,' but three specific things that you can tie to a concrete moment. Let it be as granular as possible: smell the roses on your morning walk, the smooth warmth of your coffee cup in your hands, the pattern of sunlight drifting across your desk, the voice of a friend on the phone, or the way your child made you laugh at breakfast. The more sensory detail you bring to mind, the smell, the texture, the colour, the sound, the more deeply your brain encodes these sources of gratitude. The conversation yesterday that mattered and the way the light came through the window this morning. The colleague who said the thing that helped. That is your next step, not a plan or a transformation. Tomorrow morning, try it again; the day after, do it again.

Not because gratitude practice will solve everything. Because beginning the day with your own attention and agency, directed toward what is genuinely present and valuable in your life, before handing that attention to the first incoming demand, is a small act of agency that, repeated enough times, builds something. It builds the brain's neuroplasticity, making the practice easier. It builds the morning rhythm that makes everything that follows slightly different. And it builds, over time, the experiential evidence that your attention is yours to direct, which is one of the most important things a thriving mind can know about itself.

Your brain is still being built. It will continue to be built for as long as you are alive, by whatever experiences you give it most consistently. The thriving mind you are building is not the end point, and it is a direction available to you now, in whatever small choice comes next.

EPILOGUE

The Three-Year Garden

On planting, tending, and the time it takes for things to grow.

There is a man I think about often, whom I know only from a single exchange described by Dr Mario Martinez on the Thriving Minds podcast. He was 100 years old and standing in his garden. And he turned to Dr Martinez and said, with the ease of someone reporting something straightforward: I can't wait to show you this in three years. It's going to be beautiful, a centenarian planning three years ahead.

Not in denial of mortality, not pretending he controlled outcomes. Simply planting, with the full knowledge of his age and its implications, and without apparent anxiety about either. Simply believing, in the most practical sense of believing, that there was still something worth tending, still something worth seeing, still time enough to watch it grow.

That image has stayed with me since I first heard it, and it has stayed with me through the writing of this book, because it captures something essential about what I have been trying to describe across these chapters: not a goal, or transformed self, not the achievement of some settled state called thriving, but a choice and then a decision, made repeatedly, to keep planting.

What You Have Been Planting

Every practice in this book is a seed, and every morning you begin before laying an eye on a screen, directing your attention toward what is genuinely present and valuable, before handing it to the first incoming demand, that is a seed. Every time you choose movement when stillness would be easier, every breath taken with deliberate intention to regulate rather than react, every window of focused work protected from the interruption that would have otherwise consumed it, are all seeds.

Every conversation about what AI should and should not be permitted to decide in your professional domain. Every insistence on accountability when a system produces a recommendation that your judgement tells you is wrong. Every act of mentoring that meets the standards of thoughtful practice and passes it on to the next generation is a seed.

You will not see the results immediately, because brains do not change overnight. Culture and institutions do not change overnight. Hippocampal volume does not increase in a week, and neural pathways that support sustained attention, emotional regulation, and creative genius are not built in a day's practice. They are built through the accumulation of days, in conditions that the brain can use, supported through movement, sleep, nourishment, recovery, and the repeated experience of choosing the harder, deeper, more genuinely human response when the easier, automated one was available.

That is the thing that the neuroscience of neuroplasticity establishes most firmly, and that I want to leave you with most firmly: they are built. Not in everyone, not without effort, not on a timeline you can specify in advance. But built by bodies, brains and minds that are given what they need consistently, over time.

To help you nurture your own "garden," I invite you to take a practical step: imagine a blank three-year garden timeline. On this timeline, you can plot the seeds you wish to plant, any small, consistent practices from this book or your own life that you want to see take root. These might include daily movement, protecting your first hour, naming three things you're grateful for each morning, creating tech-free evenings, or prioritising recovery. Visualising not only what you want to plant, but also when and how you expect those practices to grow, can make the slow, often invisible process of brain change feel tangible.

For example, your first few timeline entries might look something like this:

Week 1: Begin walking for five minutes each morning.

Month 2: Add a daily gratitude practice, write down three things you are grateful for every morning.

Month 3: Create a screen-free evening routine twice a week.

Month 6: Schedule one afternoon each week for dedicated recovery or reflection.

As weeks and months pass, return to your timeline to observe what has taken root, what needs further tending, and what new seeds you might add. Growth may not be immediate, but seeing your intentions mapped over time can sustain your motivation and make the long-term horizon of change feel more real. Your three-year garden begins with a single seed today.

Three years from now, if you tend these practices perfectly, but consistently enough that they become the fabric of your days, your garden will be different from the one you have now. Your baseline anxiety may be lower, and your capacity for focused thought may be higher. Your relationship with the tools of your profession, including the AI tools entering every field, may be more deliberate and more confident. Your sense of who you are, professionally and personally, may be clearer. Not because you achieved perfection. Because you kept planting.

Gardens do not bloom because the gardener resolved to grow something beautiful. They bloom because the gardener kept showing up, in the ordinary days, in the difficult seasons, in the mornings when showing up felt like too small an act to matter.

Francesca's Garden

My sister's garden never had the chance to bloom the way it should have. I have written about Francesca throughout this book with the combination of love and grief and clinical understanding that thirty-five years of thinking about what happened to her has produced. An understanding that remains incomplete, that will always remain incomplete, and that I have had to make peace with.

What I know is that her struggle planted seeds in me. Seeds that grew into questions, into a career in neuroscience, into three decades of trying to understand the brain and make that understanding useful to people navigating their own struggles. She is present in every chapter of this book; in the questions it asks and the urgency with which it asks them.

We do not always see what blooms from the seeds we plant, or from the seeds planted in us. Francesca did not know, could not have known, what her life would grow in mine. And if any of it reaches someone who needed it, who recognises something of their own experience in these pages, who finds in the neuroscience a framework that makes sense of what they are carrying, who takes one small step toward a thriving mind that they would not otherwise have taken, then something has bloomed from her life that she would, I believe, be glad of.

She changed me and through me, she may change others. That is what love does, even when it is not given the conditions it deserves.

One thing I want to say clearly before I put down my laptop is that this book is for you, regardless of how old you are, what you have already experienced, or how fixed you believe your patterns to be. The strategies and ideas I have shared throughout these pages can be adapted to different backgrounds, whether you come from a different culture, have experienced socioeconomic barriers, or are navigating life as a neurodivergent person. Your circumstances, resources, and personal history may shape which practices are most relevant or effective for you. Please know that all advice here is purposely designed to be flexible and inclusive, encouraging you to explore what works best in your own context.

The core principle is this: adaptation is not just allowed; it is essential. The most important thing is to shape these practices to fit your real life, not a prescribed ideal. Small, consistent changes, adapted to your own requirements and circumstance, are what matter most in building a thriving mind. You do not need to do everything or do it perfectly; you simply need to start where you are and take one step at a time.

To make this more concrete, imagine a few examples of adaptation: For a nurse working night shifts, building recovery into the day might mean taking short, restorative breaks between patients, even if they're not at traditional times. For someone living in a multigenerational household with little private space, a daily movement practice could be as simple as stretching beside the bed or taking a short walk after meals. A neurodivergent professional might find that traditional gratitude journaling doesn't resonate, but expressing appreciation aloud or through art does. If financial constraints are present, dietary improvements can begin with small changes, such as adding one extra serving of vegetables a day, rather than pursuing an expensive overhaul. The central idea is that the principles are adaptable: the form each practice takes can be tailored to your setting, needs, and resources.

The neuroscience of neuroplasticity does not set an expiry date. It does not conclude that the brain's capacity for change is exhausted at a particular age. An older brain requires more repetition, more recovery, and more deliberate attention to the biological conditions that support mental health. It brings advantages: greater experiential wisdom, more stable emotional regulation, and a stronger ability to see patterns and context across time. The mind at sixty or seventy is not a diminished version of the mind at thirty. It is a different instrument, with different strengths and requirements.

The gardener who plans three years ahead at one hundred is not in denial about his age. He is working with it, tending what he can plant, in the season he is in, with the understanding that even at one hundred, the garden is not finished.

Plant your three-year garden. Regardless of where you are starting from. Because three years from now, you will be three years older in any case. The question is only whether there will be something blooming when you get there.

The Invitation

This book has been an invitation. Not to perfection, but to awareness and not a dramatic transformation, but consistent practice. Not to have all the answers, but to ask better questions about the systems around you, and the technology entering your field, and of the stories you have been telling yourself about what you are capable of and what is possible.

The invitation is to see your brain not as fixed but as responsive to new beliefs, to sleep, to movement, to nourishment, to recovery, to the repeated experience of choosing the more conscious response rather than the automatic one. To see your struggles not as evidence of limitation but as the territory in which growth happens. To see the AI transition not as something happening to you but as something you are participating in shaping, through your professional standards, your ethical commitments, and your willingness to bring human wisdom to powerful but not wise systems.

The invitation is to plant and to trust the slowness of genuine change, in a world that is changing so fast that slowness can feel like failure. To understand that the thriving mind you are building, in your biology, in your habits, in your professional practice, in your relationships, is the most important contribution you can make to the world that artificial intelligence is creating around us.

When machines increase our power, we must increase our wisdom. Wisdom does not scale automatically with intelligence. It grows the way gardens grow through tending, through time, and through the quiet daily commitment to showing up.

Artificial intelligence will continue to advance long after this book is published. Models will evolve, and capabilities will expand; some of the specific examples in these pages will age. That is the nature of exponential technological change.

But the deeper principles will not age. The brain will still wire itself around what it repeatedly experiences. The nervous system will still require regulation, and institutions will still reflect the standards of the people within them. The future will still be shaped by what we repeatedly reinforce in our habits, organisations, professional communities, and in our relationships with the tools we build and use.

Technology accelerates, and human development does not, at the same pace or automatically. That is not a weakness, but it is our anchor. It is why the work of building a thriving mind is not a luxury or an afterthought in the AI era. It is the foundation of everything else.

Tomorrow Morning

Tomorrow morning, before the world arrives with its demands and its notifications and its perfectly curated feed of things designed to hold your attention whether or not they deserve it, you have a moment.

You can begin the day the way you always have, letting the patterns run while your attention is elsewhere. Or you can plant a seed. Three things you are genuinely grateful for and four slow breaths. That is enough for one morning, and then tomorrow morning, the same. And then the morning after that.

Three years from now, I hope you look back on your life and recognise something that wasn't there before, perhaps a different quality of attention, or a lower floor of baseline anxiety. A brain that has surprised you with what it turned out to be capable of, given the right conditions. A professional practice that has managed the transition with more wisdom and more confidence than you thought you had available.

I hope you look at your garden and think, look what grew from those small practices. Look what consistency produces, given time. Look at this thriving mind I have been building.

And I hope you are already planning the next three years. Because that is what thriving minds do.

They keep planting new seeds.

— The End

Notes & References

The research and evidence cited throughout this book draw on peer-reviewed studies, established neuroscience, and professional interviews on the Thriving Minds podcast. Below is a chapter-by-chapter guide to key sources and further reading.

Introduction

- The account of Francesca's experience with mental illness and the psychiatric system is drawn from the author's personal history and professional understanding of schizophreniform disorders and antipsychotic pharmacology.
- Clozapine remains the most effective antipsychotic for treatment-resistant schizophrenia, though associated with significant side effects. See: Kane J, Honigfeld G, Singer J, Meltzer H. Clozapine for the treatment-resistant schizophrenic. A double-blind comparison with chlorpromazine. Arch Gen Psychiatry. 1988 Sep;45(9):789-96. doi: 10.1001/archpsyc.1988.01800330013001.
- Neuroplasticity research referenced here draws on the work of Merzenich, MM, and Sameshima, K., Cortical plasticity and memory. Curr Opin Neurobiol. 1993 Apr;3(2):187-96. doi: 10.1016/0959-4388(93)90209-h. Merzenich, M. (2013). Soft-Wired: How the New Science of Brain Plasticity Can Change Your Life. Parnassus Publishing.
- The Adverse Childhood Experiences (ACE) study is the foundational research of: Felitti VJ, Anda RF et al., Relationship of childhood abuse and household dysfunction to many of the leading causes of death in adults. The Adverse Childhood Experiences (ACE) Study. Am J Prev Med. 1998 May;14(4):245-58. doi: 10.1016/s0749-3797(98)00017-8.

- Serve-and-return interactions and early brain development: National Scientific Council on the Developing Child (2004). *Young Children Develop in an Environment of Relationships: Working Paper No. 1.* Retrieved from https://developingchild.harvard.edu/key-concept/serve-and-return/

Chapter 1: The World Changed

- Mo Gawdat's work on artificial general intelligence: Gawdat, M. (2023). Scary Smart: The Future of Artificial Intelligence and How You Can Save Our World. Bluebird. https://www.panmacmillan.com.au/9781529077650/
- AlphaFold2 and protein structure prediction: Jumper, J., Evans, R., Pritzel, A. *et al.* Highly accurate protein structure prediction with AlphaFold. *Nature* 596, 583–589 (2021). https://doi.org/10.1038/s41586-021-03819-2. DeepMind's AlphaFold database now includes predicted structures for over 200 million proteins.
- The brain as a prediction machine: Clark A. (2013) Whatever next? Predictive brains, situated agents, and the future of cognitive science. Behav Brain Sci. 36(3):181-204. doi: 10.1017/S0140525X12000477.
- Neuroplasticity and digital environments: Loh KK, Kanai R. How Has the Internet Reshaped Human Cognition? Neuroscientist. 2016 Oct;22(5):506-20. doi: 10.1177/1073858415595005.
- Learned helplessness: Seligman ME, Maier SF. Failure to escape traumatic shock. J Exp Psychol. 1967 May;74(1):1-9. doi: 10.1037/h0024514.
- Inter-brain synchrony: Hirsch J, Zhang X, Noah JA, Ono Y. Frontal temporal and parietal systems synchronize within and across brains during live eye-to-eye contact. Neuroimage. 2017 Aug 15;157:314-330. doi: 10.1016/j.neuroimage.2017.06.018. Interview on the Thriving minds podcast #159. https://www.buzzsprout.com/367319/episodes/14306818
- Hasson et al. (2012). Brain-to-brain coupling: a mechanism for creating and sharing a social world, Trends in Cognitive Sciences, Volume 16, Issue 2, Pages 114-121, https://doi.org/10.1016/j.tics.2011.12.007.

Chapter 2: Resilient Minds

- Exerkines and brain health: Chow, L.S., et al. (2022). 'Exerkines in Health, Resilience and Disease.' Nature Reviews Endocrinology, 18, 273–289. doi:

10.1038/s41574-022-00641-2.

- Brain-derived neurotrophic factor (BDNF) and exercise: Szuhany KL, Bugatti M, Otto MW. A meta-analytic review of the effects of exercise on brain-derived neurotrophic factor. J Psychiatr Res. 2015 Jan;60:56-64. doi: 10.1016/j.jpsychires.2014.10.003.
- Associate Professor Tara Walker's research on exercise and neurogenesis at the Queensland Brain Institute, The University of Queensland. Interview on the Thriving Minds podcast episode #211, https://www.buzzsprout.com/367319/episodes/18690378
- The placebo effect and belief: Benedetti, F. (2009) Placebo effects: understanding the mechanisms in health and disease Oxford; New York: Oxford University Press.
- Dr Mario Martinez on centenarians: Martinez, M.E. (2019). The MindBody Code: How to Change the Beliefs That Limit Your Health, Longevity, and Success. Sounds True. Thriving Minds podcast episode #98, https://www.buzzsprout.com/367319/episodes/11182858
- Intergenerational transmission of stress: Yehuda R, Lehrner A. (2018) Intergenerational transmission of trauma effects: putative role of epigenetic mechanisms. World Psychiatry. 2018 Oct;17(3):243-257. doi: 10.1002/wps.20568.
- Hippocampal volume and stress: Sapolsky RM. (2000) Glucocorticoids and hippocampal atrophy in neuropsychiatric disorders. Arch Gen Psychiatry. 57(10):925-35. doi: 10.1001/archpsyc.57.10.925.
- Sleep, brain clearance, and glymphatic system: Xie, L., et al. (2013). 'Sleep Drives Metabolite Clearance from the Adult Brain.' Science, 342(6156), 373–377. doi: 10.1126/science.1241224. Plog BA, Nedergaard M. The Glymphatic System in Central Nervous System Health and Disease: Past, Present, and Future. Annu Rev Pathol. 2018 Jan 24;13:379-394. doi: 10.1146/annurev-pathol-051217-111018.

Chapter 3: The Attention War

- DataReportal (2025). Digital 2025: Global Overview Report. 5.24 billion social media users globally. https://datareportal.com/reports/digital-2025-global-overview-report
- Anxiety statistics. Australian Institute of Health and Wellbeing. National Study of Mental Health and Wellbeing. https://www.aihw.gov.au/mental-

health/overview/prevalence-and-impact-of-mental-illness

- 23 minute rule: Mark, Gloria & Gudith, Daniela & Klocke, Ulrich. (2008). The cost of interrupted work: More speed and stress. Conference on Human Factors in Computing Systems – Proceedings. 107-110. https://doi.org/10.1145/1357054.1357072
- Variable reward schedules and compulsive behaviour: Schultz W. Neuronal Reward and Decision Signals: From Theories to Data. Physiol Rev. 2015 Jul;95(3):853-951. doi: 10.1152/physrev.00023.2014.
- Flow states: Csikszentmihalyi, M. (1990). Flow: The Psychology of Optimal Experience. Harper & Row.
- Social pain and neural overlap: Eisenberger, N. (2012) The Neural Bases of Social Pain: Evidence for Shared Representations With Physical Pain. Psychosomatic Medicine 74(2):p 126-135, DOI: 10.1097/PSY.0b013e3182464dd1
- Adolescent brain development: Casey, B.J., Jones, R.M., & Hare, T.A. (2008). The adolescent brain. Ann N Y Acad Sci. 2008 Mar;1124:111-26. doi: 10.1196/annals.1440.010.
- Sextortion and online exploitation: National Center for Missing and Exploited Children 2024. NCMEC releases new sextortion data. https://www.missingkids.org/blog/2024/ncmec-releases-new-sextortion-data

Chapter 4: Your Brain Under Pressure

- Australian HR Institute (2024). Survey findings indicate that a significant majority of professionals report an accelerating pace of change in their workplace. https://www.ahri.com.au/articles/five-work-trends-to-look-out-for-in-2024
- 1 in 2 Australians Facing Workplace Burnout A community poll commissioned by Beyond Blue https://www.beyondblue.org.au/about/media/media-releases/1-in-2-Australians-Facing-Workplace-Burnout
- Burnout experiences. Maslach C, Leiter MP. Understanding the burnout experience: recent research and its implications for psychiatry. World Psychiatry. 2016 Jun;15(2):103-11. https://doi.org/10.1002/wps.20311.
- The stress response and prefrontal cortex: Arnsten, A. Stress signalling pathways that impair prefrontal cortex structure and function. *Nat Rev Neurosci* **10**, 410–422 (2009). https://doi.org/10.1038/nrn2648

- Amygdala hijack and emotional reactivity: LeDoux J. The emotional brain, fear, and the amygdala. Cell Mol Neurobiol. 2003 Oct;23(4-5):727-38. doi: 10.1023/a:1025048802629.
- Chronic stress and cognitive function: Lupien SJ, et al., Effects of stress throughout the lifespan on the brain, behaviour and cognition. Nat Rev Neurosci. 2009 Jun;10(6):434-45. doi: 10.1038/nrn2639.
- Bessel van der Kolk on trauma and the body: van der Kolk, B. (2014). The Body Keeps the Score: Brain, Mind, and Body in the Healing of Trauma. Viking. Ho JMC, et al., Book Review: The Body Keeps the Score: Brain, Mind, and Body in the Healing of Trauma. Front Psychol. 2021 Aug 18;12:704974. doi: 10.3389/fpsyg.2021.704974.
- Computations of uncertainty de Berker AO et al., Computations of uncertainty mediate acute stress responses in humans. Nat Commun 2016 Mar 29;7:10996. doi: 10.1038/ncomms10996.
- The magical number seven, plus or minus two : limits on working memory. Miller, G. A. (1956). The magical number seven, plus or minus two: Some limits on our capacity for processing information. *Psychological Review, 63*(2), 81–97. https://doi.org/10.1037/h0043158

Chapter 5: The Human Upgrade

- 150 minutes of moderate physical activity per week. Bull FC et al., World Health Organization 2020 guidelines on physical activity and sedentary behaviour. Br J Sports Med. 2020 Dec;54(24):1451-1462. doi: 10.1136/bjsports-2020-102955.
- Sleep deprivation and alcohol comparisons. Williamson AM, Feyer AM. Moderate sleep deprivation produces impairments in cognitive and motor performance equivalent to legally prescribed levels of alcohol intoxication. Occup Environ Med. 2000 Oct;57(10):649-55. doi: 10.1136/oem.57.10.649
- Associate Professor Tara Walker's research on exercise and neurogenesis at the Queensland Brain Institute, The University of Queensland. Interview on the Thriving Minds podcast episode #211, https://www.buzzsprout.com/367319/episodes/18690378
- Awe and mental and physical health. Monroy M, Keltner D. Awe as a Pathway to Mental and Physical Health. Perspect Psychol Sci. 2023 Mar;18(2):309-320. doi: 10.1177/17456916221094856.

- Professor Felice Jacka and nutritional psychiatry: Jacka FN, et al. A randomised controlled trial of dietary improvement for adults with major depression (the 'SMILES' trial). BMC Med. 2017 Jan 30;15(1):23. doi: 10.1186/s12916-017-0791-y.
- Mediterranean diet and brain health: Psaltopoulou T et al., Mediterranean diet, stroke, cognitive impairment, and depression: A meta-analysis. Ann Neurol. 2013 Oct;74(4):580-91. doi: 10.1002/ana.23944.
- Sleep architecture and memory consolidation: Walker, M.P. (2017). Why We Sleep: Unlocking the Power of Sleep and Dreams. Scribner.
- Gut-brain axis: Cryan JF, Dinan TG. Mind-altering microorganisms: the impact of the gut microbiota on brain and behaviour. Nat Rev Neurosci. 2012 Oct;13(10):701-12. doi: 10.1038/nrn3346.
- 17 hours of wakefulness and blood alcohol level of 0.05%. Williamson et al., 2000 Road Safety Research Report, Department of Transport and Regional Services Australian Transport Safety Bureau. https://www.infrastructure.gov.au/sites/default/files/migrated/roads/safety/publications/2000/pdf/Fatig_Alc.pdf
- Regular sleep schedules, National Sleep Foundation. https://www.thensf.org/setting-a-regular-sleep-schedule/

Chapter 6: The First Hour

- Morning routines and cortisol awakening response: Fries E, Dettenborn L, Kirschbaum C. The cortisol awakening response (CAR): facts and future directions. Int J Psychophysiol. 2009 Apr;72(1):67-73. doi: 10.1016/j.ijpsycho.2008.03.014.
- Breathing and the brain stem: Smith JC, et al., Pre-Bötzinger complex: a brainstem region that may generate respiratory rhythm in mammals. Science. 1991 Nov 1;254(5032):726-9. doi: 10.1126/science.1683005.
- Yackle K, et al., Breathing control center neurons that promote arousal in mice. Science. 2017 Mar 31;355(6332):1411-1415. doi: 10.1126/science.aai7984.
- Gratitude and neural pathways: Emmons, R. A., & McCullough, M. E. (2003). Counting blessings versus burdens: An experimental investigation of gratitude and subjective well-being in daily life. *Journal of Personality and Social Psychology, 84*(2), 377–389. https://doi.org/10.1037/0022-3514.84.2.377

- Habit formation and neuroplasticity: Lally, P., et al., (2010). How are habits formed: Modelling habit formation in the real world. *European Journal of Social Psychology, 40*(6), 998–1009. https://doi.org/10.1002/ejsp.674

Chapter 7: The Power Between Us.

- Inter-brain synchrony: Hirsch J, Zhang X, Noah JA, Ono Y. Frontal temporal and parietal systems synchronize within and across brains during live eye-to-eye contact. Neuroimage. 2017 Aug 15;157:314-330. doi: 10.1016/j.neuroimage.2017.06.018. Interview on the Thriving minds podcast #159. https://www.buzzsprout.com/367319/episodes/14306818
- Uri Hasson, Asif A. Ghazanfar, Bruno Galantucci, Simon Garrod, Christian Keysers, (2012) Brain-to-brain coupling: a mechanism for creating and sharing a social world, Trends in Cognitive Sciences, Volume 16, Issue 2, Pages 114-121, https://doi.org/10.1016/j.tics.2011.12.007.
- The 150 Limit (Stable Relationships). Dunbar, R. I. M. (2014). The Social Brain: Psychological Underpinnings and Implications for the Structure of Organizations: Current Directions in Psychological Science, 23(2), 109-114, http://www.jstor.org/stable/44318732

Chapter 8: The Stories We Believe

- Overestimation of driving competence: Koppel, et al., (2022). We are all less risky and more skillful than our fellow drivers: Successful replication and extension of Svenson (1981). https://doi.org/10.15626/MP.2021.2932
- Young drivers rate their driving ability as better https://rac.com.au/about-rac/media/media-releases/february-2022/survey-reveals-risk-taking-among-overconfident-young-drivers
- Confirmation bias: Nickerson, Raymond. (1998). Confirmation Bias: A Ubiquitous Phenomenon in Many Guises. Review of General Psychology. 2. 175-220, https://doi.org/10.1037/1089-2680.2.2.175
- Growth mindset: Dweck, C.S. (2006). Mindset: The New Psychology of Success. Random House.
- Cognitive dissonance: Festinger, L. (1957). A Theory of Cognitive Dissonance. Stanford University Press.

- Unconscious incompetence. Kruger, J., & Dunning, D. (1999). Unskilled and unaware of it: How difficulties in recognizing one's own incompetence lead to inflated self-assessments. *Journal of Personality and Social Psychology, 77*(6), 1121–1134. https://doi.org/10.1037/0022-3514.77.6.1121
- Four stages of learning. Martin M. Broadwell, Teaching for Learning (XVI.), in The Gospel Guardian, Vol 20, No 41, 1969. https://www.wordsfitlyspoken.org/gospel_guardian/v20/v20n41p1-3a.html
- Cognitive biases: Kahneman, D. (2011). Thinking, Fast and Slow. Farrar, Straus and Giroux.

Chapter 9: Why Are You Here?

- Purpose and longevity: Boyle PA, Barnes LL, Buchman AS, Bennett DA. Purpose in life is associated with mortality among community-dwelling older persons. Psychosom Med. 2009 Jun;71(5):574-9. doi: 10.1097/PSY.0b013e3181a5a7c0.
- Ikigai and Japanese conceptions of purpose: García, H. & Miralles, F. (2017). Ikigai: The Japanese Secret to a Long and Happy Life. Penguin Life.
- Viktor Frankl on meaning: Frankl, V.E. (1946/2006). Man's Search for Meaning. Beacon Press.
- Self-determination theory: Deci, E. L., & Ryan, R. M. (2000). The "what" and "why" of goal pursuits: Human needs and the self-determination of behavior. *Psychological Inquiry, 11*(4), 227–268. https://doi.org/10.1207/S15327965PLI1104_01

Chapter 10: Stepping Up

- AI governance and ethics: Floridi, L & Cowls, J. (2019). A Unified Framework of Five Principles for AI in Society. Harvard Data Science Review. https://hdsr.mitpress.mit.edu/pub/l0jsh9d1/release/8
- The economic dimensions of AI: Acemoglu, D, and Restrepo.P (2019). "Automation and New Tasks: How Technology Displaces and Reinstates Labor." *Journal of Economic Perspectives* 33 (2): 3–30. DOI: 10.1257/jep.33.2.3
- Professional responsibility and AI: European Commission: Directorate-

General for Communications Networks, Content and Technology and Grupa ekspertów wysokiego szczebla ds. sztucznej inteligencji, *Ethics guidelines for trustworthy AI*, Publications Office, 2019, https://data.europa.eu/doi/10.2759/346720

Chapter 11: Our Thriving Mind

- Neuroplasticity across the lifespan: Gutchess A. Plasticity of the aging brain: new directions in cognitive neuroscience. Science. 2014 Oct 31;346(6209):579-82. doi: 10.1126/science.1254604.
- Default mode network and creativity: Beaty RE, et al., Creative Cognition and Brain Network Dynamics. Trends Cogn Sci. 2016 Feb;20(2):87-95. doi: 10.1016/j.tics.2015.10.004.
- Long-term habit change and neural remodelling: Graybiel AM, Smith KS. Good habits, bad habits. Sci Am. 2014 Jun;310(6):38-43. doi: 10.1038/scientificamerican0614-38.

General References & Further Reading

- Bartlett, S.E (Ongoing). Thriving Minds Podcast and LinkedIn Newsletter.

  https://www.linkedin.com/in/professorselenabartlett/

  https://thrivingminds.buzzsprout.com/

  https://www.qut.edu.au/about/our-people/academic-profiles/selena.bartlett
- Doidge, N. (2015). The Brain's Way of Healing. Viking.
- Goleman, D. P. (1995). Emotional intelligence: Why it can matter more than IQ for character, health and lifelong achievement. New York: Bantam Books.
- Harari, Yuval Noah; Spiegel & Grau (2018). *21 Lessons for the 21st Century.*
- McGilchrist, I. (2009). *The Master and His Emissary: The Divided Brain and the Making of the Western World.* Yale University Press. http://www.jstor.org/stable/j.ctt5vkwtd
- Sapolsky, R. M. (2004). Why zebras don't get ulcers: The acclaimed guide to stress, stress-related diseases, and coping (3rd ed.). New York: Henry Holt.

- Seligman, M. E. P. (2011). *Flourish: A visionary new understanding of happiness and well-being.* Free Press.

A Note on Artificial Intelligence

This book was written in a time of rapid change. Throughout the process, I used artificial intelligence as a tool to support drafting, organising ideas, refining language, and improving clarity. It helped me test structure, challenge assumptions, and shape the flow of the writing.

What it did not do is replace authorship. The scientific insights, clinical perspectives, interpretations of research, and personal experiences shared in these pages are my own, grounded in decades of work in neuroscience and a deep engagement with the evidence.

I include this note because how we use these tools matters. Artificial intelligence will increasingly shape how we think, work, and live. The responsibility now is not just to use it, but to guide it, to train it with care, and to ensure it reflects the values we want to see in the future.

If we do this well, AI can support human potential. If we do not, it risks amplifying what we fail to question. The future is not something that arrives fully formed. It is something we are building, together, and it will reflect the attention, intention, and care we bring to it now.

About the Author

Professor Selena Bartlett

Professor Selena Bartlett is a Professor of Neuroscience at the Queensland University of Technology and an internationally recognised leader in brain health, neuroplasticity, and human resilience. Over more than three decades, she has authored over 110 scientific research papers and has been awarded the prestigious Lawrie Austin Award for Contributions to Neuroscience in recognition of her impact on the field.

Originally trained as a pharmacist, Selena retrained as a neuroscientist after seeing her sister's life struggle with a mental health diagnosis and treatment throughout her lifetime. Her research now explores how stress, adversity, and digital environments influence attention, identity, and emotional regulation in an age of accelerating technology.

Her work has helped shape national policy discussions on social media and brain development. She has testified before parliamentary committees, advised government bodies, and worked with organisations across education, law, health, and corporate sectors to strengthen resilience and performance in rapidly changing environments.

Professor Bartlett is also the creator of the Thriving Minds podcast and LinkedIn newsletter, author of six books for the general public, Miggi Matters, Smashing Mindset, Trace It to Erase It, Being Seen, Visible and now Thriving Minds in the Age of AI. She appeared in the award-winning documentary 'Seen', in the SBS documentary 'The Great Separation', and in the Queensland Police documentary 'Policing Youth Crime'. She has a regular guest segment on ABC radio promoting mental wellness strategies. Her goal is to translate cutting-edge neuroscience into practical strategies for professionals navigating uncertainty in the digital age. Her message is clear: as machines increase our power we must increase our collective wisdom.

About the Author

Made in United States
Orlando, FL
04 November 2026

90200935R00095